RAW
MATERIALS
ECONOMICS

BY CHARLES WALTERS JR.

Robert Himmerich y Valencia, Ph.D.
Major, U.S. Marine Corps, Retired

Santo Bueno Ranch
P.O. Box 1269
Peña Blanca, NM 87041

Phone: 505/438-3159
ARS WØAOM

A NORM PRIMER

RAW MATERIALS ECONOMICS

CHARLES WALTERS JR.

A NORM PRIMER

Acres U.S.A.
Kansas City, Missouri

RAW MATERIALS ECONOMICS
A NORM PRIMER

Copyright © 1991 by Charles Walters Jr.

ACRES usa Order info
P.O. Box 1690
Greeley, Colorado 80632
800-355-5313
orders@acresusa.com
www.acresusa.com

Box 95   133-9547

ISBN: 0-911311-32-7
Library of Congress Card Catalog: 91-76287

Who could have thought that it would be easier to produce by toil and skill all the most necessary or desirable commodities than it is to find consumers for them? . . . It is certain that the economic problem with which we are now confronted is not adequately solved, indeed is not solved at all, by the teachings of the textbooks, however grand may be their logic, however illustrious may be their authors.—Winston Churchill: *Amid These Storms (1932).*

*Dedicated
to the memory of
Vincent E. Rossiter, Sr.*

Table of Contents

PREFACE

A primer is a first reader.

A National Organization for Raw Materials (NORM) primer is a first eclectic reader on RAW MATERIALS ECONOMICS. Its objective is to simplify the subject so that anyone can understand it.

Accordingly, this primer contains nothing that runs beyond the comprehension of a person blessed with average intelligence and a fair education. It assumes no great knowledge of mathematics, physics, the several other sciences, or political and business economics. It may be that the reader will have to look up a few words in the glossary now and then, and handle a few new concepts—KEYNESIAN ECONOMICS, the CLASSICAL SCHOOL, GROSS NATIONAL PRODUCT, the rubrics of the *ECONOMIC INDICATORS* and the *ECONOMIC REPORT OF THE PRESIDENT*—but these things are also explained in the context of their usage. When a term used in this text is further explained in the glossary, it appears in small caps with first use, as in *Keynesian economics* above.

Having mastered the several lessons in this primer, the student of Raw Materials Economics will then be able to read the lines—and between the lines—of articles and books with superb comprehension. The issue of intellectual flim-flam will be seen for what it is, a mix of ignorance and thoroughly informed self-interest. Should a passage or chapter prove a bit difficult, the reader is urged to proceed anyway. The overview sought here will come together in any case sooner or later.

The old *McGuffey's First Reader* invoked no less than five features, each of them suggested by eminent teachers and scholars, and proceeded to teach the generations that for a century built America, often doing so in one-room schools with little equipment and a paucity of budget. It remained for schoolmen with doctor of philosophy degrees to tear the country down. "Oversimplification has become a mortal sin," wrote Clarence Ayres in *The Theory of Economic Progress* in 1944. "This is SCHOLASTICISM, the last stage in the decay of *the obvious and simple system* described by ADAM SMITH. The progress of science is always in the direction of the simplification of what seemed complex before."

McGuffey's readers relied on using small words before gradually introducing longer and more difficult ones as the pupil gained aptness in the mastery of the language. Our function here is not to teach students to read, but to think. Thus simple ideas are explained before more complex ones arrive on-scene.

This little NORM primer takes note of the fact that too often people do not understand the grammar of the subject. Some are too embarrassed to discuss the half-truths of the past and the humbug of the present. They are not sure of themselves, and have grave doubts about their ability to handle the premises of raw materials economics. Moreover, the cast of characters responsible for the most outrageous abuses of reason in economics is largely unknown, a defect we hope has been handled in our glossary.

For these several reasons, we have taken a preface out of *McGuffey's Revised First Reader*, dated June 1879, and paraphrased it for the purpose at hand.

1. Economic concepts will be introduced in a pecking order that runs from simple to complex, always preserving and explaining the rules of logic.

2. A graduation of these ideas has been preserved carefully so that each premise stands ready to support each syllogism as it is constructed.

3. The type face has been chosen carefully for unstrained reading. To preserve the spirit of the primer—even though its function is not to teach people how to read, rather how to think—the opening type is large, then graduated to a smaller size.

4. Illustrations, especially charts and graphs, have been used to convey the thought that there is nothing occult in economics, and that schoolmen hiding behind the complexities of the craft represent darkness, not illumination.

5. The primer is short, yet definitive enough to permit pupils to move on to a second reader.

Withal, a dictionary may be needed at times, albeit not an economics dictionary. Words and phrases commonly found in the latter will be defined amply in the paragraphs of this text.

For now it is enough to remind everyone that our words have a twenty-six letter base, and that base is cushioned on forty-four phonic sounds in the beautiful English language. With the words and sounds in this book, the student will be able to take on all comers when debating raw materials economics specifically, and so-called MACRO-ECONOMICS in general. This will be possible as we sweep away the esoteric baggage parked on our doorsteps by those who never tire of constructing theories, often without a single reference to the observed facts of the situation.

When you finish the primer, you should be able to explain and defend the American College Dictionary definition of *economics* as follows: "The science of treating the production, distribution, and consumption of goods and services, or the material welfare of mankind."

Now, move on to Lesson 1.

These 2 chapters are a cartoon that is as much a turn-off as it is helpful. Read The Glossary first, then ch 3 to 12 plus afterward before the comic strip, if at all.

RKHyV

MARINE CORPS HERITAGE FOUNDATION
WWW.MARINEHERITAGE.ORG

A man had it in mind one day to require his son to mow the lawn. The grass—drawing nourishment from a near perfect calcium and phosphate load in the soil—had the healthy look of a green carpet this sunny afternoon.

The boy did not want to work that afternoon, frisbee golf being the big attraction down the street a block or so. "Why can't you hire Adam next door to cut the grass?" he asked his dad.

"That's a good question," the boy's dad replied. And then he placed the usual fatherly hand on the boy's shoulder.

"If Adam mows the grass and we pay him, then we will have the grass mowed, right?"

The boy nodded agreement.

"But Adam will have our money!"

The boy's smile melted like an ice cream cone on a hot day. "Yes," the boy said, not without puzzlement. "I can see that, but . . ."

"But if *you* mow the grass, then *we* will have both the grass mowed and the money. It's a matter of simple economics."

The boy shrugged his shoulders. He started to respond, but he didn't know what economics meant. He had a boy's grasp of the fact that with verbal sleight of hand, his father had turned a personal point of view into one meant for the family. He understood for certain that he'd lost the debate because of the wry smile on his father's face.

The lawnmower motor coughed now and then as it cut swath after swath through the turf. Each time the machine hesitated in a lush clump of grass, the term *economics* came to mind, like a popular song that wouldn't leave him alone.

That evening the boy collected his fee. "Dad," he said, "will you explain economics to me?"

The older man thought about that for a long moment. He started to tell the boy about the so-called FACTORS OF PRODUCTION. How much did LABOR get? How much for RENT, INTEREST and

PROFIT? If a call to supper had not intervened, he might have lost both himself and the boy in the horse latitudes of W. W. Rostow's STAGES OF ECONOMIC GROWTH, which would have "baffled the fools and fooled the wise," just as Dreyden said.

He didn't even hint at these things because such ideas would only have compounded the mystery. Instead he told the boy about Abraham Lincoln, and reminded him of last year's vacation trip. They had stopped at Promitory Point, Utah, where the rails met the day the transcontinental railway was completed in 1869. The boy had looked with wonderment at the very spot the last spike had been driven.

"When they were building the transcontinental railway," the older man said, "some of the economists wanted to import the rails from England because they were cheaper. But Lincoln objected. He said, *If we buy the rails from England, we will have the rails, but they will have our money. If we make the rails here, we will have both the rails and the money.*

The boy thought this over. "You mean economics has to do with taking care of the family first. Or in the case of the economy, taking care of the country first?"

The answer was a soft *Yes!*

"How do you build an economy?" the boy asked.

The boy's dad thought about that a long time. Finally he said, "Have you ever read *Swiss Family Robinson* or *Robinson Crusoe?*"

With that question, both the boy and his father realized that there would be another lesson.

Lesson 1 explained an individual's own wealth creation, It, in effect, said, *Don't become dependent on others or on debt creation.*

Now to Lesson 2.

LESSON 2

How do you build an economy?

In answering that question we are required to point out that the reason government economists do not understand the exchange equation is because they always pick up an economy "in process." They have lost track of the physical connection, the link to raw materials and the role of money. These few concepts seem to confuse more than they enlighten, for which reason this primer is obliged to invoke a cartoon communication media to make everything come clear.

Suppose Robinson Crusoe's island suddenly received a shipload of people. Suppose, they, indeed, proceeded to build an economy. This idea might be presented as *THE FUTURE ADVENTURES OF ROBINSON CRUSOE*, a few panels of which were reproduced in *UNFORGIVEN* with the permission of Grosvenor Associates of London. HENRY C. CAREY, the brilliant economist who guided Abraham Lincoln, liked to refer to the Robinson same.

The Creator did not put economies on this planet. He simply gave us the earth, the sea, the sky and sunshine. Robinson Crusoe had these gifts of God—and a companion, Friday. But two people could not build much of an economy.

There was no exchange of a commercial nature. Food was harvested from nature's bounty. Clothing was a holdover from another economy. Shelter was constructed from native materials. Except for language and know-how from another civilization, Robinson Crusoe would have experienced an even more dismal existence. Read the cartoons, starting with "In Defoe's famous story. . . ."

LATER...

... AND SO, FOR MANY YEARS, FRIDAY AND I HAVE LIVED ALONE HERE.

THEN THERE IS LITTLE CHANCE OF RESCUE?

MANY OF US ARE SKILLED WORKERS... I'M A BLACKSMITH, MYSELF.

I'M A CARPENTER!

AND I, A FARMER.

As time goes on, the people get used to their government. But the barter system they have used since the last shipwreck begins to cause trouble.

As the food producer became efficient enough to feed more than his own family, a natural DIVISION OF LABOR established itself. Division of labor required exchange and equity in the exchange equation. Individuals settled this exchange for themselves as long as they were able. Before long the need for a common denominator for production values made itself felt. Thus a goat currency, which was hardly a step removed from simple barter. Next a corn currency, which Crusoe argued was standard and could be divided.

It soon became apparent that the little colony was much more sophisticated than a first glance suggested. Now that skills and people had been imported, the leaders of the island reached back to their origins for other lessons. Daring, imaginative management skills and private initiative would be needed to make the mini-civilization grow and prosper. Corn was valuable, but it lacked an enduring and solid base. In the old country a common denominator currency had come into existence—gold. Unlike corn, it never disin-

tegrated. Over-issue was impossible because new findings approximated the growth of the population in the civilized world. All nations trusted this money. But, alas and alack, there was nothing Crusoe's people would accept as tokens for exchange that did not embody value in the token itself. Then cowry shells were discovered, and a man named Arnold Benedict became a political figure.

With these thoughts in mind, let's rejoin our cartoon presentation.

In history, Robinson Crusoe remembered, gold provided a store of value and therefore a mechanism for saving for construction of a capital pool. The collapse of cowry shell money forced a return to barter and the most primitive forms of labor division.

The island's economy rotated around the total annual requirements of all the homes. Food, clothing and fuel together with outside household services, the church, schools, medicine and local government became the dominant facts of this economy. Therefore the island's economy was represented by the everyday needs of the "nation's" many households. Eventually, newly discovered gold was accepted as a common denominator for the physical input of raw materials. Manufacturers and services merely added to the price, and *enabled* a higher standard of living. Annual raw materials production—in the main, food and fuel energy—was the power gear in the economic machine and at all times it controlled the volume and velocity of industry and trade as expressed by the total national dollar income. In this equation, the integrity of the money served as the flywheel and balancewheel of the economy.

The Robinson Crusoe lessons seem clear enough. The famous loner, now joined by enough people to build an economy, identified the need for division of labor. The farmers would have to produce enough food to feed the carpenters, the blacksmiths, the teachers, and all the other professions the STATE OF THE ARTS would account for. The others would have to do something that, in effect, would make it worth the farmer's while to produce food beyond the needs of his immediate family.

State of the arts means the level of inventions. The state of the arts was not well advanced at the time man first captured fire, or when he invented his first killing tool. But by the time of *The Further Adventures of Robinson Crusoe*, say, in the mid-nineteenth century, the food producer had become efficient enough to feed more than his own family, and a natural division of labor came about. Some workers made chairs, others made farm tools, some became teachers, others brought specific chores—such as milling wheat and slaughtering animals—in off the farm. Each family gave up on the idea of doing everything for themselves. By the process of each following his own choice of work, division of labor came about.

Taxes—though very necessary—were not very popular.

Thus the rest of the story.

The island nation did not grow and prosper by clipping coupons on riskless government bonds. Pioneers built the country by mixing their labor with the raw products of the earth. It was the outward thrust of this initiative that defined the level of prosperity. But then a system was devised whereby a heavy share of the nation's income was spent as annual interest for those who had unspent income. With so much riskless income available, the incentive to take risks gradually evaporated.

As old enterprises died, new ones were not ventured. Much individual initiative was submerged by socialization of production and later by institutionalization of resultant poverty.

Arnold Benedict had his agenda. First money would be created via the printing press, later money would be based on fond hopes and fountain pen ink. There would be confusion between income generated by production and income that could be made good when answered by production. Thus we return to our cartoon scenario.

Thus, in a classical way, the citizens of Crusoe's island arrived in the modern world. The path from barter to hard money to currency and debt became littered with economic wreckage. Somehow the good people lost track of opaque reality. They forgot how income was created in the first place. And so they watched monetary authorities conjure into existence a money based on debt, not production. In so doing, these same authorities turned the old and simple exchange equation upside down.

Ultimately the confusion became so great, the citizens—almost without exception—kept their money in bankrupt institutions in order to chase an elusive interest income. They watched without understanding as they simply lost control. The forgot our Lesson 2, namely that an economy is built with production that should be distributed properly.

LESSON 3

A short piece of fiction and a cartoon have been used to flesh out the first two lessons in this raw materials economics primer. Now those same lessons must be restated in concrete terms, stripped of fiction and allusions to metaphor and allegory.

The farm wife and mother was the first economist. She was also the first manufacturer. The etymology of the term implies as much. The term ECONOMY came into the English language from two ancient Greek words, *oikos* (meaning the house), and *nomine* (meaning to deal out). Economy originally meant the art of managing a farm household. Accordingly, the woman of the house was the first farmer and the first manufacturer, a reality that gave rise to the old pioneer expression—"A man's work is from sun to sun, a woman's work is never done"—because the hunter husband had it much easier than his mate in the shelter.

Most of the nation's economy still "rotates about the total annual requirements of all the homes in our nation," wrote CHARLES B. RAY. "As of old, food, clothing, and fuel together with outside household services—the church, schools, medicine and local government—are by far the dominant facts of life in any economy. Therefore, our annual national economy is simply the sum total of everyday needs of the nation's millions of households." Even today, 70% of the purchases in the economy are accounted for by consumers—the business of food, clothing, shelter, household appliances, etc. These purchases are squeezed by conditions of unemployment and underemployment and faltering income.

Admittedly, big ships and airplanes—and military killing machines, for that matter—have glitz, as do skyscrapers and baseball stadiums, but they are no more than small punctuation marks in the economics story. *FORTUNE* 500 companies are often viewed as the key to full employment, but they in fact have not created a single new job in America in well over a decade. In fact they have shipped out some three million jobs.

Individual households and their consumption—especially food consumption—backbone the purchase activity for the entire economy. And self-employment and small businesses, farms included, supply the most jobs. When the productive system finds its locus in giant corporations and foreign climes, buying power becomes the issue.

This inability of the productive system to generate and distribute the buying power needed to consume the product has created a body of economic thought that would be ruled from fiction because of its lack of plausibility, and yet it holds whole societies in thrall. The rubrics for this captivity are many, and all seem based on the proposition that bits of activity in the marketplace somehow delivers an understanding of the whole, which they do not.

Economics, much like any science, is merely a method for "foretelling." The general method is statistical and respondent to the calculus of probabilities. The kinetic theory of gases is the best available expression of how physical laws operate, and how the concept of a scientific system discovers laws without first calibrating every detail. Any gas is composed of free molecules which are in perpetual motion. These molecules move at different speeds, drive in all directions, collide with each other and against the sides of a container. Pressure is nothing more than the measurable result of impacts, the calibration of energy of molecules striking the walls of a container. No one could possibly calculate the path of each molecule. "If by ill-luck I happened to know the laws which govern them I should be helpless," said Jules Henri Poincare. "I should be lost in endless calculations and could never supply you with an answer to your questions. Fortunately for both of us, I am completely ignorant about the matter. I can therefore supply you with an answer at once. It may seem odd; but there is something odder still, namely that my answer will be right."

Indeed, the aim of science is to foresee. Science describes phenomena and tries to join them by laws, and these laws enable man to predict.

Through the study of the motions of the heavenly bodies, for instance, astronomy has succeeded in establishing laws which enable man to calculate the position of these bodies with respect to each other in the future. In like manner, physics and chemistry describe the behavior of solid, liquid and gaseous bodies, and these descriptions lead to laws which replace the amazement of ignorance with the sureness of knowledge. When it has been experimentally observed that certain phenomena seem to be invariably linked to the first by a relation of cause to effect, this observation is worded in such a way that it enables man to predict these phenomena quantitatively or qualitatively whenever the same conditions are present.

Economics as science also is statistical. Much as with the kinetic theory of gasses, a proper overview of national economics excuses most attempts to compute the behavior of each economic molecule, each little transaction. Unfortunately, this penchant for "molecular investigation" ties up economists with endless calculations, and leaves unanswered *cause* at the overview level.

Absolute cause is rarely understood, although we are all involved in a game of *pretend*. For instance, it is a lead pipe cinch that if a truck travels a million miles, there will be an accident. That's the rule—an accident per million miles. Insurance companies have used the statistical approach ever since the

Phoenicians invented the game.

Our own probes in this primer are based on statistical evidence and findings that flow from reasoned study. There is both a limit and a demand for this approach. It defines the character of the overview and often bogs down when harnessed to an inappropriate task.

It is the function of this primer to compute the exchange equation and to point out that in order to have a national profit, raw materials must be priced in line with down-track cost factors elsewhere in the economy. But in no way can we suggest that PARITY will adjust itself in an institutional world.

We should pause for a moment and consider the term *institution*. The family is an institution, as is marriage. The church is an institution. So is the banking structure, the political party—and the list could be endless. More troublesome is the observation that these named entities are not at all alike. Yet they have a common denominator. They are all power structures.

The INSTITUTIONAL ECONOMISTS—JOHN R. COMMONS, Wesley Clair Mitchell, Thorstein Veblen—have all seen that our never ending search for a planetary system of the markets has yielded principles, and they have been understood. Even earlier, at the time of Andrew Jackson, JOHN STUART MILL wrote his scholarly *Principles of Political Economy*, a two volume treatise that surveyed the entire field of economics—rents, wages, prices, taxes. SAY'S LAW OF MARKETS still reigned supreme, as it would for a hundred years. And institutional interplay still upset the Law of Markets so regularly that to be an economist in the minds of the common folk was to be a big joke. Yet Mill did more than codify the academic thought of the hour. He brought on a principle all his own—one that has been regularly forgotten and regularly misplaced ever since. In effect, Mill argued that distribution in an institutional world had nothing to do with economics.

"The things are there," wrote Mill. "Mankind, individually or collectively, can do with them as they please. They can place them at the disposal of whomever they please, and on whatever terms. . . . Even what a person has produced by his individual toil, unaided by anyone, he cannot keep, unless by the permission of society. Not only can society take it from him but individuals could and would take it from him, if society . . . did not . . . employ and pay people for the purpose of preventing him from being disturbed of his possession. The distribution of wealth, therefore, depends on the laws and customs of society. The rules by which it is determined are what the opinions and feelings of the ruling portion of the community make them, and are very different in different ages and countries, and might be still more different, if mankind so choose. . . ."

What Mill said was transparently obvious—once it had been said. Never mind if the *natural* action of society was to depress wages, or to equalize profits or to raise rents or take farm prices down to a debilitating level. If society did not like the *natural* results of its activities, it had only to change them. Society could tax, then subsidize. It could expropriate and redistribute. It could give away all of its wealth to a king, or it could run a gigantic charity ward. It could give due heed to incentives, or it could—at its own risk—ignore them. But whatever it did, there was no correct distribution—at least none that economics had any claim to fathom. There was no appeal to laws to justify how society shared its fruits. There were only men sharing their wealth as they saw fit. It was discovery of profound consequences for it lifted the whole economic debate from the stifling realm of impersonal and inevitable law and brought it back into the arena of ethics and morality.

The raw materials economists of the 1930 went a step further. They argued that the laws of physics were of benefit to man, once understood. Men of good will could harness a mathematical mechanism to the business of economic stability, rather than let greed shaded by institutional clout run rampant.

In a way, farmers can be pardoned for not understanding the complexities of economics. Farm leaders, not trained in economics, did not discern—until it was too late—the full meaning of things like SLIDING PARITY and wealth transfer from one sector of the economy to the next.

We all know that during recent decades some schoolmen

seized the initiative and turned the exchange equation upside down. The economic advisors started their calculations at the top, with consumer price indexes. After that they subtracted wage costs, capital costs, and transportation charges from the source of production. Anything left over was assigned to the raw materials producer, including the farm operator. The process for more than four decades has not only bankrupted farmers, but also eroded the economy's markets as well. And debt creation, circa 1990s, has a hard go at creating artificial markets enough to keep the exchange equation functioning.

Dale Hathaway was Chief of the Agricultural Research Department of the Council of Economic Advisers to the President during the mid-1960s. Almost all the Washington seers of the 1960s were embracing certain tenets of CONJECTURAL ECONOMICS, tenets best stated by Hathaway in his *Problems of Progress in Agricultural Economy:* "Essentially, there are four contributions that agriculture can make to a nation's economic growth. First, it can provide the food and fiber base necessary for a population growing in numbers and in wealth. However, it is important that this be done without an increase in total resources used and/or in the relative price of farm products. In fact, economic growth is stimulated if farm prices decline so that [other] people will have more money to spend on their goods and services. Second, agriculture can provide workers to produce other goods and services by releasing them from the production of farm products. Third, agriculture can provide a market for non-farm goods and services, enabling the gainful employment of people in their production. Finally, agriculture can provide a source of capital that may be invested in improved productive facilities in [other areas] of the economy."

Hathaway was succeeded on the Council of Economic Advisers by James Bonnen. The bent of Bonnen's thinking emerged when he and Arthur Okum met with the Agriculture Committee of the Independent Bankers Association in February 1965.

"Gentlemen," Bonnen said, "if we though there was anything wrong with agriculture, we would have a plan for it. We have no plan for agriculture. . . . No, our Council has no plan for agriculture. However, there is the plan that the economists have been advocating for thirty years, but their farmers rejected it."

It was a plan that was being re-stated for the *1968 Economic Report of the President*. "Despite the revolution in agricultural technology and the attendant migration, the transformation of agriculture is not complete. The farm population will continue to decline, creating serious problems for some rural communities. The young, rather than the older farmers, will continue to be the primary migrants. This will leave behind a progressively aging population, especially among the farm poor. As a result, the natural rate of increase of the farm population will continue to fall." This was the voice of *institutional arrangements* speaking, and it was a self-serving voice.

The lack of participation in the well-being of the economy by agriculture was no accident, nor was there any compelling economic mandate for it. It was entirely due to the deliberate acts of calculating people who wanted to achieve the following:

• *The assurance of an abundance of relatively cheap and readily available farm raw materials.*

• *Comparatively cheap food, shoes, beverages, clothing and tobacco, all made from farm raw materials, to relieve the upward pressure on wages and salaries paid by large employers.*

• *A surplus labor force, amply supplemented by the influx of hundreds of thousands of dislocated rural employables annually.*

• *A comparatively docile union labor force, properly chastened by an abundance of willing non-union workers who are capable of innocently, or maliciously, disrupting the traditionally unionized labor force.*

• *A farm raw material price level that can be manipulated from 100% of fair value to less than 50% of fair value so a profit can be made on both the purchase and again on the sale.*

These were the real accomplishments of the farm programs since 1952, and what they intended to accomplish.

It is ironic that the school door opened opportunities for rural America, only to have the land grant college school door annihilate its clientele. This institution was organized to protect. Schoolmen, more than farmers, had a clear vision of where they were taking the raw materials producers. For instance, it was Michigan's professor Kenneth Boulding who wrote. . . . *"Civilization is what happens in cities, and the city is dependent on there being a surplus from the food producer and on some existing organization which can take it away from him. With this food surplus, the political organization feeds kings, priests, armies, architects and builders, and the city comes into being. Political science in its earliest form is the knowledge of how to take the food surplus away from the food producer without giving him very much in return."*

Few have either answered or refuted Boulding. Indeed, Henry Cantwell Wallace may have been among the last to hold up a different vision.

Early in the summer of 1924 Secretary of Agriculture Henry Cantwell Wallace gave expression to a philosophy, one that had transported a new people almost eons beyond what other men in other lands had been able to accomplish. Secretary Wallace died before he was able to finish *Our Debt and Duty to the Farmer*. With Nils Olsen, Henry Agard Wallace completed the work in 1925. Either Henry A. or his associate wrote that "men of vision must arise soon if the United States is to be saved from the fate of becoming a preponderantly industrial nation in which there is not a relation of equality between agriculture and industry. They must act in the faith that it will be good for the entire nation if agriculture from henceforth advances on terms of absolute equality with industry. They must ever keep before the mind of the nation the long-time point of view both materially and spiritually. They must set the minds of farmers on fire with the desire for a rural civilization carrying sufficient economic satisfaction, beauty and culture to offset completely the lure of the city."

Later, when Henry A. Wallace became Secretary of Agriculture during the depth of the depression, he named Mordecai Ezekial as economic advisor. "Factoryward ho!" screamed

Ezekial, and it became a watchword, a program for agriculture, and finally a public policy. The circle between international trade, industrial power and government sanction of class exploitation started closing. There remained only a suitable way of moving toward the goals without sparking revolt. Sliding parity proved to be the tool. It permitted policy makers to pace the rate of farm bankruptcy, to keep it high enough and low enough so that farmers would drown without making too much noise?

For generations, economists have been trying to calculate the movements of micro-miniature economic molecules, locked in a game of pretend. The rubrics they chant have been SUPPLY AND DEMAND, LAISSEZ FAIRE, FREE TRADE, COMPARATIVE ADVANTAGE, the IRON LAW OF WAGES, SAVINGS EQUAL INVESTMENTS, the PROPENSITY TO CONSUME, the MARGINAL EFFICIENCY OF CAPITAL, the rate of interest, on and on, all pseudo-measurements of individual importance, yet incapable of answering the question Winston Churchill posed as early as 1932, but which still cries out for an answer (see page *v*).

Lesson 3 says that economies must develop laws based on science that answer the distribution puzzle. Perhaps knowing this, historian and philosopher Robert Heilbruner wrote about the *unbearable anguish* he and his colleges would feel if they "imagined ourselves as executioners of mankind."

LESSON 4

We now come to an analysis of what happens when division of labor takes hold, and production is enhanced by improvement in the state of the arts. The business of providing food, clothing and shelter came first, and it was handled by the first laborer and capitalist, the farmer, as we have noted. Thus the starting point in the American economy—between 1700 and 1800—was the farm. Farmer pioneers created the first capital and therefore the first wealth of the nation by using raw labor, the capital of the individual, and harvesting the raw materials of the earth. Fully 90% of the nation's population lived in rural areas at that time. Manufacturing and service industries were insignificant during the era represented by the cartoon story in our second lesson.

This primitive economy was essentially a one-wheel economy. The arrival of the steam engine, other inventions and improved farm implements accounted for a two-wheel economy at some uncertain point before the Civil War. This meant that the farmer could now produce enough to feed himself and a spin-off worker in a service profession and the emerging factory system. From that moment on the wherewithal to consume the production of the economy was dependent on the mechanical law of the lever as a guiding principle. This economic equation was first stated by RICHARD CANTILLON in 1735, and was adopted by the middle of the 18th century by the PHYSIO-CRATS of France. Elements of the idea emerged in the writings of W. STANLEY JEVONS and also in Adam Smith's *An Inquiry into the Nature and Causes of the Wealth of Nations*, circa 1775.

It is not all that difficult to handle the abstraction we are obliged to consult. Simply stated, the farmer hires and pays himself. In 1800 a farmer, through his basic labor and capital and production hired only himself and a fraction of another worker. By the time of the Civil War, he hired himself and another worker, not via the legal convention of ownership and payroll, but in terms of an economic function. By 1927 this same composite producer of raw materials hired himself and four other workers—providing only that these raw materials producers were paid 100% annual raw material price parity, which represents the real value of labor and capital involved considered economically.

The raw materials producer may therefore be said to hire all other labor in society in accordance with an immutable economic law. This law requires an exchange equation based on full employment and operating capacity, fueled by parity in every sector of the economy.

Thus it may seem strange to hear the Chicago Board of Trade boast that it is there "to discover prices for commodities." Most recently this has meant that when the openly worshipped market forces took 1989 soybean prices too high for short contract holders, CBOT discovered lower prices that took between two and three dollars per bushel out of the coffers of nearly a half million soybean producers. CBOT, so far, has not discovered the relationship between the monetary value of raw commodity production and earned national income. Nor have the discovery experts been able to compute the non-value of NATIONAL INCOME based on credit injections not governed by profits and savings.

If the institutions that discover price, and the schoolmen who chart it endlessly for a meaning, were both honest and competent at the task, they would by now have discerned the consequences that attend any economic short circuit of buying power due to both labor and capital in agriculture. They would have discovered how fluctuating and speculative futures markets deprive the national and world economy when they cause production units, times price, to fall below par exchange. This discovery of prices suitable to certain business institutions causes poverty, insecurity, famines, revolution and wars. This may be the reason Berthold Brecht, the author of *Three-Penny Opera*, concluded that famines are *created* by the grain trade.

In short, physical facts, not abstractions preside over the exchange equation. This means that agriculture is the flywheel and balancewheel of the economy. It is a leading indicator because its price movements lead the economy by a bare minimum of at least six months. In fact it can be said that agriculture leads the rest of the economy by even more, because it takes years for an infant to grow up and enter the labor force. Agriculture—because it is the nation's largest industry, and because it creates the biggest share of the raw materials input of new wealth in any accounting period—must be considered the foundation of the economy of the United States.

Our eclectic first reader approach makes it a required activity to visit with CARL H. WILKEN, a founding father of the old Raw Materials National Council, the predecessor of National Organization for Raw Materials.

Agriculture, the industry which creates a continuous flow of new wealth from the soil, is the foundation of the economy of the Untied States. From the production of raw materials, all human and animal life obtains the energy to perform the labor basic to all production and economic progress. Agriculture's raw materials furnish the physical production for labor. The units of agricultural production times the price per unit give us the gross agricultural income. This in turn governs the total national income, and consequently the income of all segments of our economy.

The earned income of other segments of our economy follows the trend of gross agricultural income with almost *mathematical precision*, changing as efficiency and new products replace the old. In spite of the various fluctuations, however, the mathematical ratio of the other segments of our economy to agriculture can be traced with an accuracy comparable to that of other sciences. In short, our economy is a science based on agriculture, which is the key to economic welfare and prosperity.

Benjamin Franklin, one of the founders of the American system, [writing in *Positions to be Examined Concerning National Wealth*, April 4, 1769] pointed out that there were three ways in which a nation might become wealthy:

• By war, which permits taking by force the wealth of other nations.

• By trade, which to be profitable requires cheating. For example, if we give and receive an equal amount of goods and services through trade, there is no profit other than that obtained in our own production cycle.

• By agriculture, through which we plant the seeds and create new wealth as if by a miracle.

The records of our economy bear out the wisdom of Franklin's analysis. There may be those who, because of their academic environment and fixation of thought resulting form conventional economic teachings, will not agree that agriculture is the governing factor in our economy. Our advice to them is to read this analysis taken from the record of the nation. We are not trying to present some theory of economics. The record speaks for itself.

It has been constructed because our form of government gave rise to the best economic system the world has yet seen. In fact, with an abundant supply of mineral and agricultural raw materials, we have the basis for an almost complete economy within our borders. For the few items which we need to add to our production, we are capable of producing surpluses which the world will gladly accept in trade.

In order to guide the thinking of those who read this analysis, we will state the premises—as we find them—that say agriculture is the governing factor in our national economy.

First, we have a capital economy. Agriculture's capital investment is two-thirds that of all productive enterprise. To short-change the income of agriculture through low prices for its products is in reality to force a depression, de facto or real. In comparison with other segments, for example, agriculture has over ten times the capital investment of the steel and automobile industries combined. Further, it employs ten times as much labor as do the two industries mentioned, if functions formerly performed on the farm—slaughtering, processing, etc.—are considered.

Agriculture furnishes most of the raw material income of the nation. Other raw materials account the remainder of the raw material income in an almost perfect ratio of two parts gross agricultural income to one part gross mineral production (including coal, petroleum, copper, iron, bauxite, phosphates, potash, etc.).

Records constructed since 1910-1914 reveal that an increase in agricultural income precedes the rise of factory payrolls and the income of other segments of the economy.

Agriculture, because of it peculiar situation as an industry dependent upon climate and other natural conditions, is always in full production with approximate full employment. The production factor changes only as governing weather conditions change. The reason for this is quite simple. A farmer pays full tax on his acreage whether he owns or rents his productive factory. Not having any control over production, it is only natural that he plants his full acreage, subject, of course, to a policy of sound agronomy which will permit the most production.

Agriculture is the foundation for the production of raw materials which enter into non-durable manufactures, or items of manufactured products which supply the essential for human existence—chiefly food, clothing, and household operation.

Agriculture must feed and nurture the future labor force for human progress. Hence, it becomes necessary for agriculture to feed the increase of population up to the point that they become available for economic growth or expansion. Agriculture, therefore, precedes the rest of our economy by at least sixteen years.

The most important reason why agriculture is the governing factor is seated in the livestock population. This insight is of ancient origin.

In thinking of ancient economists and philosophers, the symbol of wealth was cattle. More wealth could be attained only through ownership of more cattle, and in the scheme of things this led to plowed ground and crops so more feed for more cattle could be produced.

Farm animals mean free labor. Millions of head of cattle, plus millions of hogs, sheep and poultry—obeying nature's primeval urge to survive, increase and multiply—function as an incredible labor force, gathering crops and processing them into usable protein for the human race. It has been estimated that half the farm land in the United States couldn't even be harvested without livestock. Without these economic production units there would be no return from the unimproved and non-tillable farm land.

In addition to gathering the crop on half the farm land, these animals consume the grass and tame hay grown on improved farm land, plus approximately 85% of the feed grains. This animal inventory must be viewed as a production machine that creates new wealth without wages or strikes. Added to the labor capital of the farmer himself, this joint force becomes the dominant labor force in the nation.

These conclusions are not the visions of the theorist. They flow from the facts, and they argue that the facts can be read in no other way.

An analysis of the record from 1850 to 1920—seventy years of growth—reveals that employment on the farm and in the factory changed not at all—one person on the farm being balanced by one person in the factory. This constant ratio of employment on the farm to employment in the industrial sector was merely a reflection of the use of raw materials as the agricultural economy expanded its production and farm functions migrated to the cities. Needless to say, without raw materials from the farms, the industrial sector could not have been built.

During the time frame 1850 to 1920, agriculture expanded by adding five million acres of improved land each year. Since then the number of harvested acres has become the plaything for programs and theories that equate shortage with price maintenance. Some of the production claims made are preposterous. For now it is enough to note that between 1920 and 1944 the average yield per acre for nine storable commodities—wheat, rye, buckwheat, corn, oats, barley, cotton, flax and hay—was approximately the same as during the twenty-five years from 1895 to 1920. More recent production record will be examined once we dispose of the proposition that faltering production can use the agency of supply and to maintain either farm or national income under conditions of full employment, or the second economic conceit, that competition at near world prices can become the fix America needs to achieve stability.

Having survived the Great Depression, the world seemed as one yawning chasm, awaiting a new line of thinking. Instead it got John Maynard Keynes and his opinion—never supported by facts—that the real cause of unemployment was a lack of investment. It got Alvin Hansen and his blessing on the Say-RICAR-

DO school, which he pronounced "fundamentally sound," and his hex on the LAUDERDALE-MALTHUS-SISMONDI advocates, whose ideas, he said, were "logically untenable." Unemployment, said the great professor Hansen, was a consequence of imbalance between wages demanded by labor, interest commanded by money lenders, and prices at retail. Never once did the post WWII school take a hard look at whether the index constructed by wages and capital cost was used to loop back to raw materials the proper price level that would be needed to support first the sales, then the profits and savings required to deal with the first fact of economic life—people and a satisfactory way of life for people. It will be noted on Table 6, page 99, that the average gross savings for the twenty-five years ended in 1953—at which time parity was annihilated—were approximately 95% of the total value of all raw materials used in operating the economy.

In 1934, British economist CHRISTOPHER HOLLIS wrote that "the historian has to record that in almost every age there was some superstition or other of utter unreason which strangely occupied the minds of men," namely the mystical significance of numbers, the claims of astrology, the study of gizzards of birds, and the strange superstition that whenever money is created, a percentage must be paid ever afterwards as a propitiation to a banker. Included with the above might as well have been the chant à la Say, *supply is demand, general overproduction is impossible*, and the pontifical intelligence out of Harvard: "Supply of one thing is a demand for another," and "Demand as a whole can never be less than supply as a whole." S. M. MacVane in fact made those statements without a single reference to the observed facts of the situation.

What, indeed, are the observed facts of the situation?

In Mexico and Spain, they talk romantically of the "moment of truth." The moment of truth is that split second when the bull is killed. If the bull dies gracefully, having put up a good fight, there is great applause. If the final moment becomes a melee, and the bull splatters blood and excrement on the suit of lights, or perhaps impales the matador, then the event is said to be graceless and lacking in finesse. It really doesn't matter whether the bull fights or not, not from the bull's point of view. He's finished the moment he's selected for the day's event. The bull must die. It doesn't matter whether he fights well or badly, he will die. Higher forces have decreed it.

Something of this nature was ordered for America's family farmers hard on the heels of WWII, in 1946. Advisers in the wings of politics said, *Empty the countryside*. And they knew how to do it.

These advisers came on as The American Farm Economics Association, the COMMITTEE FOR ECONOMIC DEVELOPMENT and the COUNCIL ON FOREIGN RELATIONS. The American Farm Economics Association surfaced in 1945 with a contest styled, *A Price Policy for Agriculture, Consistent with Economic Progress, that will Promote Adequate and More Stable Income for Farmers*. Of 317 papers entered, 100% agreed that the parity base period, 1910-1914, was null and void, and that farmers ought to have a go at free international trade again, with government money keeping the worst features from showing up domestically.

All of the papers seemed to argue that the market could not consume full production from all the well-managed farms of America, and that therefore the concept of parity represented a cultural pantheon, an outmoded article of faith. Many years later, Jimmye Hillman of Arizona University (as President of the renamed American Agricultural Economics Association) said that in place of parity, policy makers must consider an income concept which "will demonstrate the actual income of farmers, including current income and changes in net worth," including the opportunity costs of the resources of farmers in uses outside of farming.

These few sentences and paragraphs have been written for this primer to illustrate the scope of the debased pragmatism operative under the color of the economic discipline. Most of this thinking became classroom, then public policy fare because those licensed to think about economics correctly sensed the wishes of "a few industrialists who wanted to export manufactured goods in exchange for cheap farm products," wrote Carl H. Wilken with the advice and consent of ex-Tariff Commission head JOHN LEE COULTER, and Sears-Roebuck engineer Charles B. Ray. "These industrialists either failed to realize—or they didn't care about the consequences—that by exchanging industrial production for cheap farm commodities already available in the United States they created a cheap market in other nations. At the same time they reduced their domestic market to the cheap level of foreign markets."

The farm problem has never been one of aggregate overproduction. Except for seasonal variations, production has generally matched population growth. But when price per unit (bushel, carton, bin, whatever) has been made to fade by the price discoverers, farm income then started to short circuit the TRADE TURN needed to service factory payrolls. Up to the year the public policy managers discovered debt injection as a permanent way of life, farm raw material income meant a factory income in approximately the same amount. Farm and factory then locked arms to deliver a national income on a ratio of one unit of farm income to seven units of national income. But the moment the great exchanges discovered a lower price, the shortfall in farm income triggered a similar shortfall in factory payrolls and a fall-off of national income as governed by the efficiency ratio.

These data do not represent conjectural economics. Rather, they represent simple arithmetic. They annihilate the longstanding "finding" that the farmer grew so much stuff it couldn't be consumed if we stayed up all night. We will defer to a later lesson the oft-heard surplus charge, which is usually based on a wheat and cotton inventory without a single computation in evidence to factor in the displacement values of imports of everything from alcohol to red meat and poultry.

This fourth lesson of our little primer can profit by the inclusion of a few paragraphs from Carl Wilken's 1947 book, *Prosperity Unlimited*.

The simple facts are that our problem was not a question of over-production but one of *underconsumption*. There is a serious question as to our ability to produce enough for our nation, if we maintain farm prices at parity with the consumption this would bring about. The world is hungry and ill clad. We can exchange our few surpluses for those things we do not have or of which we do not produce enough. If we are to have world trade, we must produce a surplus of some things in order that we may have goods to trade. It might as well be cotton or wheat as some other product. The problem is to handle this surplus without impoverishing the cotton farmer, and to use the surplus to get full value from other nations. We might point out in this connection that if we sell our surpluses at world price levels and buy back at world price levels we always get full value, regardless of price. But cheap foreign prices should not be imported into the U.S. as demanded by unregulated trade.

With gross farm income the governing factor, and with agriculture always in full production, weather permitting, the *earned* national income has to be a multiple of the gross farm income. The laws of physics require it. We will examine these laws in more detail in the lessons that follow.

This much said, we can now state principles of raw materials economics that will be supported in the following pages.

1. Fully 70% of all economic activity is geared to answering the wants of the population—food, clothing and household comforts.

2. Farm production at parity governs *earned* national income, and therefore its maintenance supports national income on an *earned* basis.

3. A 5% increase in domestic buying power is generally equal to the benefits of all international trade.

4. When exports as a percent of national income exceed 6.5%, the trade process presides over out-migration of companies and jobs to foreign climes.

5. Raw material inputs in operating the economy determine the number of jobs in fabrication and distribution.

6. The turnover of primary income from raw materials into national income has a ratio fixed by the state of the arts (in the absence of excess debt) when producing consumer goods.

LESSON 5

The laws of physics and chemistry provide matchless insight into economics if we have the wit to see. Albert Einstein once put it this way: "God doesn't play dice with the universe." It was his conception that order, not chaos, was in charge at the highest level, and that it could be in charge at any level if the laws of the Creator's management were obeyed.

Thus there are physical laws and theorems—the laws of gravity and electromagnetism, Pascal's theorem, Archimedes principle, the law of the lever, on and on. So far we have merely hinted at the rules of arithmetic that require monetization of raw materials entering a production cycle if there are to be profits and savings at the other end. Schoolmen have scoffed at the idea on grounds that chaos, not order, is in charge. But science says differently.

Science says *all energy comes from the sun*, which is more than an axiom. It is the foundation for life.

Two or more photons make an electron.

Two or more electrons make an atom.

Two or more atoms make an element.

Two or more elements make a compound.

Two or more compounds make a substance.

Two or more substances make a cell.

Two or more cells make an organ.

Two or more organs make a system.

Two or more systems make up bodies—human, animal or plant.

But long before we consider crop production and animal and human life, we are required to scrutinize the elements of the earth and their warehoused condition as compounds.

It was Dmitri I. Mendeleyeff who first constructed a table of known elements in their natural order, and supplied the periodicity of property and weight. His insight into the Creator's order was so great he supplied spaces where he believed an element should be to comply with the rhyme and reason of that supreme plan. That table still stands. The blank spots have been filled in. And the Periodic Chart of Elements has provided a simple and beautiful picture of order in the universe where we live. It did more. It opened chemistry and physics as never before, and made it possible for lesser minds to understand the structure of the atom.

Here, for instance, are typical Periodic Chart of Elements entries for hydrogen, nitrogen, phosphorus and potassium (kalium).

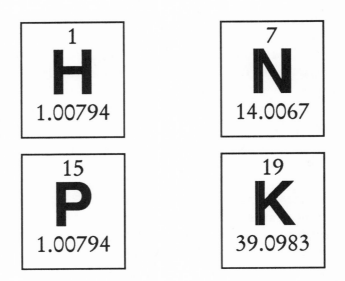

All the elements needed for life are listed as the first fifty-three of ninety-two natural elements on planet earth. Of these, all except one falls in order among the first forty-two, and all except two are listed among the first thirty-four. There is also a natural order for abundance of elements, according to atomic weight and number. The heaviest elements are the rarest. Elements with even atomic numbers are more abundant than those with odd numbers in our universe. We don't know why, nor can we even guess.

The table itself is a veritable encyclopedia. There are series with missing electrons. As the eye moves from titanium to zinc, unfilled orbits change, an electron at a time. These transitions take place in natural order, moving across the table. There is also a vertical order to the table, weight increasing as each element is listed under the one above. There are groups that figure in biology and signal the entrance or exit of disease. A heavier metal can displace a lighter one in the same group in biological tissues and alter the reaction of the lighter one. Tissues with an affinity for a certain element have an affinity for all other elements of the same group. Some elements in two groups are liver and kidney seekers.

These few data only hint at the vast complexity of the Creator's inventory plan. They also hint at how little of this vast knowledge has so far been uncovered, even though nature has been revealing herself for a long time.

Our scientific exchange equation is based on the Table of Elements, now lifted to raw materials status. There are between forty and fifty-five world commodities that determine the income of the world in general, and the income of any nation in particular. These raw materials are listed in groups across the next page, and weighted according to the best available data on value in the world's annual production cycle. Lenders may write on a piece of paper, "10,000 hogs," but not even a child tracing letters in a *McGuffey's First Reader* would be dense enough to believe that this "wealth creation" *exercise* actually produced 10,000 hogs, nor would it cause residual wealth to trickle down, as the investment school suggests.

TABLE 1

20	19	9	12	11	11	18
Wheat	Cattle	Cottonseed oil	Cotton	Platinum	Coal	Hides
Corn	Sheep	Olive oil	Wool	Gold	Coke	Newsprint
Oats	Hogs	Coffee	Silk	Silver	Natural Gas	Lumber
Soybeans	Poultry	Cocoa	Hemp	Titanium	Petroleum	Turpentine
Barley	Lard	Tea	Jute	Magnesium		Linseed oil
Rye	Butter	Sugar		Bauxite		Phosphates
Rice				Cobalt		Potash
				Nickel		Nitrate of soda
				Chrome		Rubber
				Manganese		Linseed meal
				Uranium		
				Scrap iron		
				Pig iron		
				Zinc		
				Tin		
				Lead		
				Copper		

This list could be fleshed out to include spices, fish, opium, tobacco and precious stones. The last raw material of significance to be added to the international trades has been blood, and—with the advent of abortions worldwide—human fetuses.

The number over each column designates the approximate weighting of each raw materials group in terms of world production. Many of these commodities are traded on the exchanges in Chicago, where pit workers—wise beyond their years—discover their prices. Or they are traded via satellite with international brokers taking checkoff profits, handling the inadequate remainder to the primary producers of wealth.

This inventory of raw materials has been included here simply to make the point that raw materials, not bank notes, start the production cycle, and that capital is the result, not the cause, of technology inserted into the production sequence. These raw products are the foundation of world income, and those produced in the United States are the real foundation for our own economy. In terms of the entire world, Third World nations included, most countries are more dependent on farm income than the United States. Indeed, it is incumbent on each nation to feed itself, otherwise it will never develop capital (via production times price) with profits and savings approximately equal to monetization of raw materials in the first place.

Income for any economy is created or destroyed by the raw materials intake and the monetization thereof. This means that the cost factors must loop back to furnish a statistical base for the raw materials price level. This price level governs not only national income on an *earned* basis, but also the level of profits and savings.

To prove this statement we are required to turn to the economic record. An audit of that record proves that during three five year periods of full employment and stability since 1910, farm raw materials accounted for approximately 70% of all raw materials involved in running the economy.

The first five year period was 1910-1914, which became the base year "100" for computing farm parity with the arrival of the first parity laws.

The second year period was 1925-1929. This became the select base period for Office of Price Administration (OPA) price ceilings during WWII.

The third such period was 1946-1950, a post-war era during which price stabilization for farm raw materials was operative as a consequence of the WWII stabilization measure still in force.

Any of the above base periods can be used to stabilize the income earned by farm raw materials. There have been no such periods since 1950 simply because the discovery was made that debt could synthesize national income—the consequences of investment beyond the level of savings to be handed to future generations.

The remaining 30% of raw materials used to operate the economy during the three full employment periods mentioned here can be characterized basically as mineral raw materials. Included are the metals—gold, silver, copper, lead, zinc, iron, manganese, tungsten, iron, scrap iron, etc. Fuels are a big component. Included are coal, oil, natural gas, alcohol. All are power energy sources and help supplement human and animal energy in effecting production.

So-called non-metallic inputs include bauxite, uranium, rock, sand and gravel, cement, phosphates, and potash and calcium for agriculture.

No more than forty to fifty storable raw materials—including basic storable farm commodities—construct the foundation for the full economic cycle. The volume of these commodities times price times the trade turn defines and limits the national income on an *earned* basis.

Professional economists have had difficulty handling this abstraction. They can't seem to identify the source of the money, and the mechanism for its creation eludes them. Yet the problem is as simple as the answer to the small boy who moved the lawn so that the family could have both the lawn mowed and the money.

If a sourdough finds a gold nugget, and turns it over to the government for money, the nugget costs nothing, even though the miner walks off with a fist full of dollars.

This same principle was invoked by Benjamin Graham in *Storage and Stability*. Graham wanted a commodity dollar that could circulate the same as a silver dollar, a Federal Reserve Note, a United States Note or a National Bank Note. Under his plan some twenty basic commodities—everything from steel ingots to wheat—were to be supported by the government "The only expense connected with instituting and maintaining the monetary storage system is the cost of storing the various commodities in the unit," Graham argued. "The actual acquisition of the commodities does not involve any expense or entail any annual interest charges, for they pay for themselves by qualifying as the backing for currency in the same way as our gold and silver reserves have always done." This in fact became the Coulter-Ray-Wilken parity approach, namely the monetization of basic raw materials.

Wilken proved with government statistics that raw materials production had a threefold effect:

1. The raw materials supply determines the number of jobs available in fabrication, processing and use—from raw materials production to manufactured products—and distribution.

2. The dollar value put on this new wealth raw materials production determines the amount of money which can and must be used to produce, buy and move through the economy the same raw materials production. As various costs are added—chiefly labor and capital costs—the add-on factors pyramid themselves into national income. The turnover of primary income from raw materials into national income for the twenty-five year periods, 1929-1953, meant five dollars national income for each one dollar of raw materials utilized in producing consumer goods.

3. The value placed on raw materials automatically becomes the initial market for the exchange of manufactured goods. During these several periods of full employment and controlled debt expansion, fully half of the market for manufactured goods remained in rural America, where it was created via the production and income from raw materials.

A scientific exchange equation must be based on the law of the lever, regardless of the psychological impulses schoolmen insert into the formula. It must seek out a norm, which in this instance is simply the first fact of economic life—people, a satisfactory way of life, full employment and prosperity. That is why raw materials economics shuns the debased pragmatism and bleating demands of the merchant for *more*.

Our system is statistical. It relies on bodies of data that we with our finite minds think lead us to cause. All causation is multiple. Causes always function in integrated patterns and configurations. They come from the whole of which the event is a part. And the scientific system has taken it upon itself to distill the common denominators from the body of the whole so that proximate causes can be discovered. As a matter of fact, gathering economic data has been a prime chore in the United States government since before the days of George Washington.

All the great American economists—Henry Carey, FREDERICK LIST, WILLIAM M. GOUGE—were mightily concerned with finding the relationship between general principles and stubborn facts, namely the facts of simple arithmetic.

Others have pondered the problem.

When Thorstein Veblen detailed his concept of evolutionary capitalism, he added insight to the riddle of economic life. When John R. Commons took on Veblen's teachings to become a social reformer and public administrator, he dealt with symptoms and signs and avoided construction of a theory. But when Wesley Clair Mitchell resolved to test each hypothesis against stubborn arithmetic, he uncorked the beginning of the answer to the booming, buzzing confusion. The three became known as institutional economists. They became the intellectual descendents of Richard Cantillon, Francois Quesnáy and the Physiocrats. They reasoned with a fine singleness of purpose, and they taught their students lessons that the institutional *powers that be* had a hard time putting down.

One student out of the Wisconsin of John R. Commons was John Lee Coulter, later Dean of North Dakota A&M, later a member of the U.S. Tariff Commission, still later an associate of Carl H. Wilken. The information flow between Coulter and Wilken drew constant nourishment from Veblen, Commons and Wesley Clair Mitchell.

In the process of writing a *History of the Greenbacks*, Mitchell observed the paucity of good statistical information. His career thus became a scholar's search for statistical techniques. Working on the problem Mitchell visualized a work entitled, *The Money Economy*, but when it was published in 1913 it went forth as *Business Cycles*. Hard on the heels of WWI, Mitchell gave direction to the business of statistics gathering at the National Bureau of Economic Research. Today, the finest set of statistics in the history of man—those gathered by the U.S. government—can correctly be counted as a monument to the career of Wesley Clair Mitchell. The income equation used by Wilken might never have been developed without the data gathering apparatus Mitchell and his associates accounted for.

Although it wasn't accomplished in twenty-four hours, those early researchers finally set up the elements for an income equation for the American economy. In short, the total national income had to be the sum of the component parts. A cost to one segment was income to another, and vice versa.

It did not take an astute seer to observe the relationships. The gushing well that kept the system moving was the private enterprise system. Individual capitalists save and capitalize businesses or farms. They pay out the wages and capital costs to do a job, and recover from the market if they are successful.

By the early 1930s, SIMON KUZNETS—later awarded a Nobel Prize in economics—had developed the *Economic Indicators*. This fabulous array of statistics invoked pure logic, and set a long range goal of structural balance for the economy. Kuznets reasoned that national income was made up of six component parts. All forms of income finally telescope until they fall under one of these heads. The total of all six becomes the national income.

The six segments are:
1. Compensation of employees (wages).
2. Income of unincorporated enterprises.
3. Income of farm operators (NET FARM INCOME).
4. Corporate profits.
5. Net interest.

6. Rental income of persons.

The *Economic Indicators* publication became the backbone for the *Economic Report of the President*, which was mandated by the Employment Act of 1946.

Depending on adjustments in the state of the arts, each component in the exchange equation has to earn its share of the national income if the exchange equation is to function without short-circuits, excessive debt and finally economic convulsion. This is not theory speaking, but mathematical science as exact as equations used to put a man on the moon.

Kuznets further reasoned that the income sectors had to perform well enough to pay the costs, or the economy would slowly and steadily sink into the misery and convulsion that seems generally to be waiting in the wings.

Economic Indicators require us to be able to handle certain abstractions. The publication tell us that the welfare of nine money center banks and perhaps one hundred *Fortune* type military-industrial-university complex companies do not mean the welfare of the country.

The *Economic Indicators*, if read properly, support the reasoning that national income is no unfathomable happenstance. Available data iterate and reiterate the fact that the average price level of all goods and services constructs the consumer price level. Therefore the units of goods and services multiplied by the price level at which exchange takes place determines the national income. The national income created by production of goods and services represents the effective demand, or ability to buy the annual production. Without production there can be no supply to multiply times price. That is why physical production and price per unit for that production are equally important. No production times whatever price equals no income. Full production times any price shortfall becomes the glue that holds under-consumption together.

There are several conceits that must be stated, then explained, before the exchange equation comes clear. The first is the idea that surpluses cannot be sold except through lower prices. Yet price reduction at the raw materials level and at any stage of transfer—fabrication, services and sales—destroys the income needed to consume the production. Mathematical ambition will always be vanquished by physical possibility. Lesson 6 will make that point come clear.

LESSON 6

The raw materials exchange equation invites proofs more detailed than those so far presented. Engineers who compute the weight of water behind a dam, or the nautical requirements for keeping a ship afloat, can have confidence in the rules of calculus and mathematics. The chemistry of a material and the stress factors computed in the structure, after all, are based on physical laws, and not on abstractions such as the dollars conjured into existence by a banker on the basis of collateral already on hand.

This is the reason we have included Table 2 on the following page. There is no reason to shun a table, chart or graph. In a way, these devices communicate more clearly than narrative, and leave no room to squirm for those who reject the premises of our earlier primer lessons. Nevertheless, it must be pointed out that the several columns presented have been codified from more detailed statistical arrays. For instance, the column styled *nonmetallic/metallic* includes totals for copper, zinc, lead, manganese, tungsten, gold, silver, iron and scrap iron production.

The column for *fuel* includes crude oil, natural gas, bituminous and anthracite coal.

TABLE 2
VALUE OF MINERAL PRODUCTION
IN THE UNITED STATES
(Millions of Dollars)

YEAR	NONMETALLIC METALLIC	FUELS	OTHERS	GRAND TOTAL	REALIZED GROSS FARM INCOME
1929	1,480	3,190	1,217	5,888	13,832
1930	986	2,765	1,015	4,765	11,420
1931	570	1,892	704	3,167	8,378
1932	286	1,743	432	2,462	6,400
1933	417	1,683	455	2,555	7,050
1934	549	2,233	543	3,325	8,465
1935	733	2,330	587	3,650	9,585
1936	1,082	2,759	716	4,557	10,627
1937	1,468	3,201	745	5,413	11,185
1938	893	2,820	650	4,363	10,037
1939	1,292	2,834	788	4,914	10,426
1940	1,679	3,117	819	5,614	10,920
1941	2,132	3,708	1,038	6,878	13,707
1942	2,364	4,103	1,109	7,576	18,592
1943	2,488	4,608	976	8,072	22,870
1944	2,340	5,178	899	8,417	24,113
1945	1,975	5,212	954	8,141	25,323
1946	1,825	5,760	1,311	8,896	28,967
1947	2,909	7,941	1,634	12,484	34,002
1948	3,510	10,362	1,894	15,766	34,520
1949	3,050	8,723	1,875	13,648	31,763
1950	3,701	9,506	2,205	15,412	32,066
1951	4,380	10,695	2,505	17,880	36,944
1952	4,214	10,449	2,600	17,263	36,842
*1953	4,683	10,992	2,975	18,650	35,430

Source of Data—U.S. Bureau of Mines for mineral production.
* Government payments included, starting in 1953. These data cover the classic twenty-five year period before debt construction became the engine used to drive the economy.

The column labeled *others* includes bauxite, uranium, quarried stone, sand and gravel, cement materials, ag lime, phosphates and potash.

Note that during the twenty-five year period from 1929 to 1953, fully 70% of the raw materials input used to operate the economy was of farm origin. We will consider this twenty-five year time frame as a model, for it will be noted that after 1952-1953, the nation's managers presumed to repeal the laws of physics, and then injected debt no longer justified by profits and savings as a substitute for earnings.

Note that on Table 3 full use has been made of the preliminary data assembled in Table 2. The last column of Table 2 has become the first column of Table 3. Here again, emphasis has been put on the time frame 1929-1953, with commentary on later years deferred for down-track analysis. A glance back at Table 2 will reveal that realized gross farm income now resurfaces as column 2 in Table 3. The total of the minerals in Table 2 is now column 3 in Table 3. Gross farm income and gross mineral income are added together to become column 4, which serves up the total for mineral and farm income.

To get gross farm production as a percent of national income, divide national income into gross farm income. The total of the twenty-five years national income so divided into farm income yield a rounded 14.2%. Taking the entire period, here is the arithmetic.

$$3{,}476{,}400 \overline{\smash{\big)}\ 493{,}464.000}^{\ 14.19469}$$

The quotient is 14.19469, or 14.2 rounded off.

To get gross mineral production as a percent of national income, divide national income into gross mineral production. Again the twenty-five year period divided into national income delivers the figure to be 6.04% of national income.

The combined total of raw materials production—farm and mineral—divided into national income provides an answer of

TABLE 3
TOTAL VALUE OF RAW MATERIAL PRODUCTION
AND PERCENTAGE OF NATIONAL INCOME
(Billions of Dollars)

YEAR	NATIONAL	GROSS FARM	GROSS MINERAL	TOTAL MINERAL & FARM
1929	87.8	13,832	5,888	19.7
1930	75.7	11,420	4,765	16.2
1931	59.7	8,378	3,167	11.5
1932	42.5	6,400	2,462	8.8
1933	40.2	7,050	2,555	9.5
1934	48.9	8,465	3,325	11.8
1935	57.0	9,585	3,650	13.2
1936	64.9	10,627	4,557	15.2
1937	73.6	11,185	5,413	16.6
1938	67.8	10,037	4,363	14.4
1939	72.7	10,426	4,914	15.3
1940	81.6	10,920	5,614	16.5
1941	104.7	13,707	6,878	20.6
1942	137.7	18,592	7,576	27.2
1943	170.3	22,870	8,072	30.9
1944	182.6	24,113	8,417	32.5
1945	181.2	25,323	8,141	33.5
1946	179.6	28,967	8,896	37.8
1947	197.2	34,002	12,484	46.5
1948	221.6	34,520	15,766	50.3
1949	216.2	31,763	13,648	45.4
1950	240.0	32,066	15,412	47.5
1951	277.0	36,944	17,880	54.8
1952	290.9	36,842	17,263	54.1
1953	305.0	35,430	18,650	54.0
	$3,476.4	493,464	209,756	703.8

These data—ended the year excess debt construction became public policy—gave the Raw Materials National Council its one to five ratio, meaning $1 of raw materials times the trade turn multiplier delivered $5 national income on an earned basis.

20.24508%, which rounds off to 20.25%. That is where we get
the equation that says gross national income will be five times
more than the gross of the raw materials input under conditions of parity prices and full employment. Only a change in
the state of the arts which advances agriculture beyond its
twenty-five year relationship with the industrial sector could
upset that natural formula.

$$3,476,400 \overline{\smash{\big)}\ 703,800.} \quad .2024508$$

The raw materials exchange equation that rachets national income to a level five fold higher than raw materials income is
also based on the proposition that without debt injection to
alter the equation—fully 70% of all economic activity is geared
to answering the wants of the population—food, clothing, and
household comforts. The statistical support for this finding follows on the next page as Table 4.

We must reflect on what these data are telling us, and we are
required to permit the results to be developed from the
evidence. So-called common sense and/or theories will not
serve us in this endeavor simply because too few laymen will
invoke the proper level of abstraction, and too many profes-
sionals will attempt their task with unclean hands. The idol
called science has become the servant of vested interests. A
military-industrial-university complex has enlisted at least 70%
of all scientists for the purpose of developing more sophisticated killing machines. On the basis of debt this complex has
lifted apparent gross national product to dizzy heights. High-
ways and cities—usually constructed on the basis of debt—daz-
zle the populace, and opulence created by transferring wealth
from one sector to the next has conspired to deny the prime
mover status of the raw materials on which all industrial
development is based.

Review the three charts presented in this lesson, then proceed to Lesson 7.

TABLE 4
COMPARISON OF CONSUMER EXPENDITURES FOR FOOD, CLOTHING, BEVERAGES AND TOBACCO TO TOTAL CONSUMER GOODS EXPENDITURES
(Billions of Dollars)

YEAR	I Consumer Goods Expenditures	II Food, Beverage, Tobacco	III Clothing	IV Columns II & III	V Percentage of Column IV To Column I
1929	46.9	21.2	11.2	33.4	70
1930	41.2	19.4	9.7	29.1	70
1931	34.4	16.2	8.2	24.4	71
1932	26.4	12.7	6.0	18.7	71
1933	25.7	12.8	5.41	8.1	70
1934	30.9	15.5	6.6	22.1	71
1935	34.4	17.6	7.0	24.6	71
1936	39.1	20.0	7.7	27.6	70
1937	42.1	21.6	8.1	29.6	70
1938	39.7	20.6	8.02	8.7	72
1939	41.8	20.9	8.4	29.3	70
1940	45.0	22.2	8.9	31.1	70
1941	52.9	25.7	10.5	36.3	69
1942	58.3	31.2	13.1	44.2	76
1943	65.8	36.4	16.0	52.4	80
1944	72.1	40.1	17.5	57.6	80
1945	81.2	44.6	19.7	64.3	80
1946	100.4	52.3	22.2	74.5	74
1947	113.7	58.0	23.0	81.0	72
1948	120.9	61.4	23.9	85.3	71
1949	120.4	60.7	22.7	83.4	70
1950	129.0	63.2	22.7	85.9	67.3
1951	138.2	71.2	24.2	95.5	70.0
1952	142.8	75.1	24.8	99.9	70.0
1953	148.7	77.2	24.6	101.7	68.4

Source of Data—U.S. Department of Commerce.

These data—developed for the 1929-1953 twenty-five year classic period—reveal that 70% of all consumer expenditures have to do with household and family maintenance.

LESSON 7

Any eclectic reader must restate its premises routinely before moving on to higher ground. Carl H. Wilken did this with his many publications and with the figures furnished by the government's data collectors.

"The American people have become a group of specialists," said Wilken, "and have forgotten that each group is interwoven with every other group in an indivisible economy, with each group a multiple of the complete economy of the United States. As special groups gain advantage over each other, they immediately find that other sectors of our economy—those which furnish the markets—do not keep pace with them in the consumption of goods, and they all fall back into what is usually called a depression. In fact, a depression is nothing more than an unequal price balance between groups. With an ample supply of raw materials and labor, proper pricing of goods and services should automatically create the income to exchange and consume our production."

Here, in edited abstract form, are Wilken's foundation concepts.

It is a simple fact that the financial measure of our economic welfare, whether individual, corporate or governmental, consists of adding up two columns of figures—income and disbursements. Regardless of what our theories may be, these two totals tell the story of our economic well-being.

Income consists of *primary bartering power*, which is created by the production and sale of new wealth—things obtained from the earth, farms, mines and seas—and earned income, which is derived from wages, interest and profits.

Disbursements include everything on the "outgo" side of the ledger, whether in the accounts of an individual or the government. Even the wages and salaries of those in public service must be regarded as disbursements, since public employees are not producers of wealth.

The amount of primary bartering power, or primary income, depends upon two things—the number of units produced and the price received for them by the producer. In the processing industries and professions the amount of income is governed by hours or days of labor times the rate of pay. It is therefore fundamentally necessary that the total annual production of goods and services rendered, times prices, plus wages, interest, fees and profits must create an income large enough to pay for all the costs of operating the nation as a business. The total must pay for the costs of government, pay the cost of producing raw materials and for their processing and distribution.

Our problem has not been one of production, but of distribution. We have been unable to get the necessary distribution of income, which in turn is ability to buy, into the hands of consumers. Our problem, then, is to put more dollars in the hands of the consuming public. This can be done only by increasing the number of units of production and by maintaining proper prices for the goods produced and of course by maintaining wages, fees, profits, interest rates, all of which supply income and in turn purchasing power.

How, then, could money be withdrawn from the capital reservoir soundly? There was only one way, Wilken argued. "The sound method of drawing money from the capital reservoir is by the annual production and sale of goods and service." In other words—by earning it.

Wilken used a bushel of corn as an example. When the primary producer, the farmer, took the corn to market, the first stop in the distribution system became a fact. The elevator operator was equipped with capital that had been created through the years of an expanding economy, and the savings of people. With his capital he helped make up, and in fact was a part of, the reservoir of credit dollars. When the price of corn was eighty cents a bushel, the elevator operator drew eighty cents out of the capital reservoir and paid the farmer for the corn, and to him the eighty cents constituted new money, money that offset the bushel of corn now part of the economy. The eighty cents did not have to be repaid by the farmer.

But, when the price of corn was only forty cents a bushel, the elevator operator drew only about forty cents from the capital reservoir, "and simple arithmetic tells us that the flow of money from the reservoir in that case is just half as large as when the corn is priced at eighty cents.

Unfortunately, businessmen think as businessmen must. Their psychology tells them that profits are predatory, and not participation in the welfare of the whole. Thus, the ritualistic attack on parity and the banker's love affair with debt. Using economic mythology as a smoke screen, the equation is made to appear as a charade wherein producers have merely to present collateral security and pay interest to lenders. These lenders then create money out of thin air so that borrowers can produce at the expense of the entire community. There is a problem with all this. No money is created to make distribution possible. Obviously it is as important to supply money to consumers so that they can consume as it is to create money so producers can produce. Obviously, also, the business of larding out new money always into the production side of the equation tends to cause consumption to falter.

It is this business of creating pure smoke money for one sector of the economy, and denying earnings to another sector, that has established the principle that inventory buildup results in faltering parity, when in fact faltering farm parity results in inventory buildup.

Suffice it to say that with barter, commodities of equal value change hands. This concept was preserved in precious metal money. The principle underlying the use of gold and silver was perfectly in line with the principles of modern physics. It holds that real wealth cannot be created out of nothing, but must be harvested and manufactured from the raw materials and energy sources of the globe and sun. Therefore no individual or institution should be permitted to create a claim to wealth out of thin air, for each buyer ought to give up something of value equal to that which is acquired.

So much for this lesson. Astute students will capture the main points in Lesson 7 and write them on their slates or notepads. The glossary will furnish in-depth explanations of economic concepts, always in small caps when introduced.

Mature reflection on these several points will prompt the pupil to return to physical basics. Just because new wealth may be obtained for capital, wrote Frederick Soddy in *Wealth, Virtual Wealth and Debt*, "by getting some other individual who wants it to take it in exchange, this must not blind us to the fact that a nation cannot turn its capital wealth back into consumable wealth again, or eat its plows if it is short of bread."

The great Alfred Marshall defined economics as the study of how a man gets his income and how he uses it," and he made a distinction between "consumers' goods" and "producers' goods," but his nomenclature was at least as vague as it was of practical value to the economists who saw the world only in terms of how the individual got his income. Not so when the focal point of the explanation is the nation. The physicist Soddy fairly chaffed at the reality that so few understood the difference.

Arising, however, from this difference of viewpoint, the orthodox economists seem to have committed a definite error of accounting which vitiates their whole effort to account for the monetary system, and why it is behaving so erratically and spasmodically. When one passes from the conception of wealth as a "realized amount" to the more elegant conception of it as a "periodical receipt" or flow—and on the energy theory also one is, of course, really dealing with flows—we must not omit to account correctly for what may be termed the wealth in the pipes, meaning the total amount of partially produced wealth in existence corresponding with any given rate of delivery or revenue ("volume of trade"). Thus the great American oil industry [in 1930] uses 100,000 miles of pipelines, which permanently hold three-quarters of an American billion (1,000 million) gallons of oil. The *quantity* of three-quarters of a billion gallons of oil has to be put in, but does not come out, though the oil does. We may say this quantity of oil is not burnable, though the oil is—that, though the oil is always passing through from production to combustion, three-quarters of a billion gallons are as good as wasted so long as the supply is maintained. . . .

If this quantity is not honestly accounted for by someone abstaining from consumption to an equivalent extent, it accounts for itself dishonestly, by the something-for-nothing money trick, and lowers the value of every ones's money by changing the value of each unit.

In our time, great exchanges and a subservient government have emptied the pipeline to near collapse simply to keep alive the confusion between the individual interest and the national interest. And this confusion has been used to annihilate the par exchange that alone can sustain full employment and preside over the structural balance necessary for maintenance of full employment. Reread this lesson again, and you will be ready for Lesson 8.

LESSON 8

The last few observations in Lesson 7 bring us back to where we started—to the elements and raw materials of the earth, and to parity. The Founding Fathers could well have answered, *Why parity?*, had the question been put to them in its modern context.

Subsection 5 of Section 8 of the U.S. Constitution was adopted in order to establish for "ourselves and our posterity" a par economy. This subsection authorizes Congress "to coin the money, regulate the value thereof, and of foreign coin, and to fix standards of weights and measures."

The third act of the First Congress was a tariff law to prevent cheap foreign goods and debased foreign currencies from determining the value of American money. It has always been the government's job to regulate that value, and failure to regulate that value today constitutes gross dereliction of duty by all those in the House, the Senate, and the President as well. And the value of money cannot be regulated without par exchange for the raw materials of the earth.

The powers that be—the international business houses—have always believed low raw material costs in one land and high markets in another constituted the royal road to greatest profits. These same houses have always relied on a great spread between costs and sales domestically. Few have realized that business principles are not the same as principles governing an economy, and fewer still now realize that principles governing an economy ultimately govern business. In the U.S., business tried to circumvent the par economy with cheap labor from Europe and the Orient, and finally labor got an immigration law. International business tried to circumvent the American cost level with cheap imports, and in the 1920s Congress passed the FORDNEY-McCUMBER tariff bill so that cheap imports could not rupture the American price structure.

There was a farm bloc in those days, and political pressure was brought to bear so that a McNARY-HAUGEN bill to preserve farm parity could be passed. It got a veto from Calvin Coolidge. Finally, policies from 1890 to 1930 gave us our Great Depression. It was in the era of the 1920s and 1930s that great thinkers in Congress gave birth to the parity concept, the idea that agricultural production must exchange on par with the rest of the economy.

It was understood then that parity is a requirement of any economy based on division of labor. To understand this, reason with us for a moment. If two individuals exchange the production accounted for by division of labor, and one has a 10% advantage in each transaction, the favored party will have most of the money or property, and the second will be near bankrupt or bankrupt after the tenth transaction.

This is what has been happening? This is what the fight for farm parity was about in the 1920s and 1930s. In the late 1930s, Congress did order the parity concept computed. From the very first, the USDA operated so as to confuse and discredit the parity concept. Because of ignorance, or thoroughly informed self interest, government functionaries started computing parity for over one hundred fifty commodities one at a time. Parity was questioned routinely by government economists. As a result the Farm Act of 1938 put a 70% ceiling on parity for farmers if the crop year delivered 70% of a so-called normal crop. Since production times price equals income,

70% of a full production times 70% of a parity price meant 49% of a proper income level for agriculture (70 x 70 = 49). The economists didn't seem to understand that 70% of an income for agriculture meant 70% of employment for industry. And so the depression continued, and the farmer was told about the rubrics of supply and demand.

By 1932, farm income had dropped over 50% from the 1929 level. Even a short crop was without a market because income was not generated to consume the production. What about supply and demand? There is no supply and demand. There is only an interrelatedness between purchases and sales on a short term. The demand is, in fact, created by supply if the supply is priced at par in the first place. As a matter of fact, of the twenty-six grain crops produced between 1909 and 1934, the thirteen largest sold for more per bushel than the thirteen smallest. Production has not outrun population expansion and there has in fact been no surplus—except seasonal—since 1910-1914. Nor can there be under a condition of par exchange unless there is no hunger, no poverty or no possibility for farm raw material participation in plastics, fuel, medicinal and other industrial production.

Howard W. Hjort, President Carter's Director of Economics, Policy Analysis and Budget, USDA, once prepared an analysis on parity. This was fed to senators and representatives, to then USDA Secretary Bob Bergland, to deputy secretaries and the many assistants in those many acres of bureaucracy on the Potomac. Hjort told one and all that *relying on a sixty year old reference point presents problems. Farmers no longer purchase the exact same items they did in 1910-1914*, he said. They had horses then, an oats economy, worse yet—no electricity, no hybrid seeds or oil company fertilizers or chemicals.

It strains the imagination to understand what the age of a measuring device has to do with the accuracy of that device. Parity, after all, is a ruler, a yardstick. The years 1910-1914 were selected as a parity base period because they represented a time bracket when the dollar stood at "100" for every sector of the economy. The parity price level is the level at which there is the greatest distribution of goods. It is the key to distribution. It was also the point of our maximum need for foreign goods. True, agriculture had those dirt roads then, those awkward headers, no electricity, but industry was equally primitive. Technology is like a giant milk tank. Add to the pool of technology, and it flows evenly to all parts of the tank. As the flow into an economy improves, more people are released into service industries—teaching, research, government—but these service industries cannot be sustained without parity for agricultural raw materials unless there is unsound debt expansion.

Computation of parity depends on little more than a good base year—one in which there is no import invasion, no general imbalance in the economy—a year in which basic storable farm commodities are at par with wages and capital costs.

As pointed out earlier, the 1910-1914 era was a fair base period. The 1946-1950 period computed by Carl Wilken could have provided an equally valid base period "100." Computations on the price of corn and wheat will deliver the same price within a penny figured on either base period.

But something happened in 1946—something that put American agriculture on a collision course with bankruptcy. Harry S Truman was president then. He had inherited the Stability Act of 1942 and the Steagall Amendment. This legislation had the effect of putting a 90% of parity price on most storable farm crops at harvest. Parity of 90% at harvest on storable commodities turned out to be very close to 100% with storage added, and storage is properly the farmer's function. There was reasoning behind this stabilization measure. Senator Elmer Thomas of Oklahoma stated it best when he reported back on a farm bill for parity—one geared to stabilizing the dollar—and asked the measure be assigned to the Committee on Banking and Currency, its logical base for consideration. After all, he said, it was the job of Congress to regulate the value of money. But there was a war on then, and the establishment was in danger. When Truman took over, the war was nearly over. In 1946 he could and did take no less than four steps that haunt us now.

1. He signed a measure to make permanent the temporary withholding act of WWII. This assured us all that we could expect continued growth of government and unending funds for bureau expansion and regimentation of men, resources and capital.

2. He supported the Administrative Procedures Act. This took off the hands of Congress the unhandy business of writing laws in detail. They could now write enabling legislation and hand off to bureaucrats the right to issue laws by decree, not just by the page, but by the pound.

3. He supported and signed the Employment Act of 1946. This made it public policy to secure full employment for industrial America, largely at the expense of agriculture.

4. He did—on the last day of 1946—declare the war terminated. The Stabilization Act had been written so that it would expire two years after the end of the war, whether ended by Congress or by presidential proclamation.

The Employment Act of 1946 set up a Council of Economic Advisers. Edwin Nourse was the first chief, and he was replaced by his deputy, Leon Keyserling. At first there was a little talk of structural balance in the *Economic Report of the President*, but soon this waned, and the language became the language of policy papers out of Committee for Economic Development, the Council on Foreign Relations, and pronouncements out of what Theodore White called New York's Perfumed Stockade.

The year 1948 arrived all too soon. There was great debate in Congress that year, and at 5:00 a.m. on the morning the Republican Convention met in Philadelphia to nominate Thomas Dewey, conferees reported out a farm bill. It contained a strange clause, one that escaped much attention until recently. The 1948 Act provided for base periods to be moved forward every decade. When 1957-1959 became base year "100"— farmers lost forty-nine cents off a bushel of corn with a lead pencil in the computations. The base then became 1947-1949 equals "100;" next 1957-1959 equals "100." Base periods moved forward each decade with no references whatsoever to the requirements of a good base year.

You can date the declining rate of profit for industry from the date farmers experienced the loss of farm parity. You can watch liquidity slide right out of the banks as of that date. Former USDA Secretary Charles Brannan's plan for a relief check instead of a price for agriculture didn't help, except to pace the rate of farm bankruptcy so it wouldn't happen too fast for political comfort.

This declining rate of profit has been accommodated by inflation these many years, and by unions trying to keep their members by asking for inflationary increases, and by business enterprise accommodating the unions despite the fact that industry did not have the ability to pay the asking price.

This inventory of circumstances has had its consequences. In order to cover up a lack of par exchange between different sectors of the economy, government has expanded its role. Our rule of law or common devotion to precedent is being abandoned by the courts and overridden by legislative action in the passing of statutory law. Bureau people bestow benefits on those shared out by a lack of par exchange, but they do so in exchange for votes, for campaign funds, for favors, for job offers, for bureau expansion—all of which yield benefits indirectly to some at a cost to others. Stability of private rights is by its very nature a constraint on what government can do. To the extent that government power to revoke or abrogate rights is limited, the market for the services of individuals in government is limited.

It can be seen that non-agricultural raw materials are fairly successful in maintaining parity. Natural gas, for instance is regulated. Service industries—the telephone and other utilities—in effect have price control. Iron and steel at least have powerful corporations to shelter them from the ravages of speculative markets, albeit not from foreign invasion. This leaves the kind of commodities listed on the exchanges to the vagaries of speculation and manipulation. Most of those commodities are agricultural, and these commodities are the backbone of that 70% expenditure known as sales to households.

We are thus obliged to ask questions and explain answers having to do with parity.

There is a lot of criticism concerning parity because it is old and based on 1910-1914 equals "100." This criticism is not valid. There is usually agreement that 1910-1914 was a golden age for agriculture, that there was relative balance between the price of what a farmer sold and what he bought. But look at how things have changed! Look at agriculture in 1910-1914— dirt roads, a horse and oats economy, jolt wagons, few tractors, no milking machines, awkward headers, few combines, no electricity. Well, this is true, but these things have very little if anything to do with the point in question. Parity is a yardstick, a measuring device. The foot is sometimes dated back to King John, sometimes to the ancients who built the pyramids. Obviously a measuring device does not suffer because of its age if indeed it does a good job.

Does parity as a measuring device do a good job? Of course, if the premises are understood. Unfortunately these are not well understood by most of the press, politicians and Congressional leaders—certainly not by USDA if we accept what they are doing as being intellectually honest. The governing factor is the base period. It has to be properly selected, and this requires correct economic information and an understanding of logic. You can find one of the best rationales in Thorold Rogers' *History of Agriculture and Prices*.

Parity as a measuring device made a transition to the American scene when the Founding Fathers were active. Benjamin Franklin understood it. If you want to look at it that way, the Revolutionary War was really a monetary war. England was exploiting raw materials and delivering manufactured goods at a price differential. The colonies did not like this. They smuggled. They evaded the law. One of the final straws was when the Bank of England forced on the colonies British pounds. Up to that time the locals coined their own money based more or less on commodities, or warehouse receipts. But in the modern context, parity got a toehold in the American language when George Peek wrote *Equality for*

Agriculture shortly after WWI.

George Peek called it a "fair exchange value." In other words, the American economy was no longer a one wheel economy. It now had plenty of division of labor, and continued prosperity under a condition of division of labor depended on an appropriate division of income between the several sectors of the economy. The buying power of a crop should be on par with the buying power of factory labor, services, professions, taking these things in terms of broad spectrum averages. Without this balanced exchange equation, a shortage of buying power would sooner or later short circuit the system. That's why they went back to 1910-1914 equals "100." There is an assumption here. And that assumption is that technology development will benefit the several sectors of an economy about the same. True, they had jolt wagons in 1910, but the factory wasn't computerized either. In the 1920s and 1930s, many factories still ran with a central power plant and lots of pulleys and belts running all over the place. Picture it now—an automobile rolling off the assembly line every few minutes. All these arguments against the parity measuring stick suggest that farm technology has outrun factory technology in efficiency, and therefore farmers ought to constantly express their efficiency advantage in terms of lower prices. This is fiction.

George Peek came up with the following calculations. In terms of a 1910-1914 index, corn sold at 64.2 cents a bushel. Wheat came to 88.4 cents a bushel. To compute true parity prices for any year, all one had to do was take the 1910-1914 figures and multiply times an average of the things a farmer had to buy. The government has commodity indexes, processed food indexes and the like. We know how these have increased. Simply take that figure and multiply. For instance, it is now generally agreed that the big bull market in stocks peaked in 1966 in terms of stable dollars. At that precise point the commodity index stood at 240%. Thus the multiplier became 340 for that year since you have to add the base period of "100." This meant that corn parity for that year should have read this way: 64.2 cents x 340% = $2.18. Wheat: 88.4 x 340% = $3.00. Total income in 1910-1914 averaged $33 billion per year. The industrial wage was twenty-three cents an hour. It took several hours of wages to buy the equivalent of a bushel of wheat in those days. It took about twenty-five minutes to pay for that bushel in terms of the national wage average. At a $10.00 national wage average, a bushel of wheat would be worth approximately ten minutes, illustrating that the individual wage often is also below parity.

This seems simple enough, assuming that we have accurate figures on how the rest of the economy has fared. Would it be possible to move the base period a little closer to modern times—just to satisfy the doubters?

The answer is *yes*, provided, however, that a year was selected in which there was relative balance in the exchange equation between rural and industrial America. The year 1926 could be used as a fair base period equal to "100." An even better one is 1946-1950 equals "100." During the 1946-1950 period, farm prices averaged—all things considered—at 99.5% parity. Corn averaged $1.54 a bushel and wheat averaged $2.01 a bushel. If you follow through on what we've illustrated, you'll find that the consumer price index adjusted 43% upward between 1946-1950 and 1966, the point of reference cited earlier. So if you multiply $1.54 corn times 143% (the 43% increase plus the base of "100"), then you get $2.20 corn. Wheat on the same general basis comes to $2.87. This is a penny or two different from a 1910-1914 projection, but that difference certainly is not staggering.

One may ask, this is the way it works, what's the big hang up about using this yardstick for farm prices? The answer is not as simple as it might be. The parity idea started getting clouded in the very beginning. Take the tobacco crop. They decided not to use the 1910-1914 base period because it was considered out of line. So right off they used a base period of 1919-1929. This could be questioned. Then there were special concessions for soybeans. Here was a crop all but unknown in 1910-1914. In 1937 the pre-war base for milk was routinely questioned. As a result Congress gave the Secretary of Agriculture the right to decide milk prices in federal order markets. In 1940 they changed the base period for tobacco to 1935-1939, to depression years, if you will, and all this kept the parity equation dancing around stage center in a chorus of question marks. By 1944 USDA was computing parity figures for some one hundred fifty-seven commodities. No more than sixty-one were on the 1910-1914 base period. The sixty-one, however, did account for 82% of the farm crop values. At that time seventy-three commodities had a 1919-1929 base, twenty-one made use of various combinations drawn from the 1920s, and two used 1935-1939. The simple yardstick wasn't so simple the way the government did it.

Yet it is simple if you keep a sound base "100" in mind. Under the War Stabilization Act of 1942 and the Steagall Amendment, farm prices were maintained at relative parity for approximately ten year by the market. There were war years and there were years of peace involved. This means it was economic stupidity that kept the United States in a depression during the 1930s. The administration told the public that there were farm surpluses. Henry Wallace even burned the pigs and corn to short the supply. At the same time we imported more than we exported during each of the depression years. We imported more than we exported during the war years—just to catch up. The War Stabilization Act and the Steagall Amendment were written so that 90% parity for farmers would end two years after the war. Truman took the step of ending the war on the last day of 1946.

Why 90% parity? Why not 100%? It was reasoned that the farmer produces a year in advance. It is his economic function to store the crop inventory. At 90% of parity at harvest, with storage added, it turned out to be very close to 100% of parity.

Truman ended the war in 1946. And this was the reason there was a farm bill problem in 1948. The Republicans met in convention in Philadelphia that year. At 5:00 a.m. on the day the convention started, Congress finally came up with a farm bill. It continued the 90% parity bill for one year, and also enacted the 60 to 90% of parity idea into law as soon as that one year was gone. Truman signed this legislation. The decision had been made that agriculture would henceforth be the shock absorber for the rest of the economy. When the 1948 law was passed for agriculture with a 60 to 90% of parity law provision—it contained a mandate that the parity base year would be moved forward every decade. So you had this situation: 1947-49 might be a fair base period—relative balance in much of the economy! But a decade later 1957-59 equals "100" became the base period because of this law. Under the earlier base period, corn was $2.04. When 1957-1959 became the updated parity base year, they simply reduced corn to $1.55 and called it full parity. They took forty-nine cents off a bushel of corn this way.

The parity base period has been moved since then—to 1967-1969, for instance, and to 1977-1979, and 1987-1989. Each time it has been moved the legislated base period has mandated a new "100." This means a move of the base period in effect rigged the figures and called whatever the prices were for that period "100," even though they might have been quite lower. They do this after the fact. This is the reason farm prices are expressed at 77% of parity or 80% of parity when in fact, in terms of honest parity, they're much more under par with the rest of the economy.

The year 1962 produced a relatively honest array of statistics. The next year compilers of the *Economic Report of the President* changed many of the figures for every year clear back to 1929. This has been going on ever since. One of the tables in that book carries the all-commodities index, another has the farm products index and still another contains the processed foods index. Using the 1962 report, figures for 1947-1949 are arrayed at the top of the next page.

Obviously prices were not in balance when the first move was made to the base period in compliance with the 1948 farm act. You can see farm prices were 26.9 points lower than all commodity prices, and 16.0 points lower than processed foods expressed as an index.

YEAR	ALL COMMODITIES	FARM PRODUCTS	PROCESSED FOODS
1947	96.4	100.0	98.2
1948	104.4	107.3	106.1
1949	99.2	92.8	95.7
Total	300.0	300.0	300.0
Average	100.0	100.0	100.0

By dividing the three years, one can see that they average "100," suggesting a very balanced base period. Now using this same source of reference, let's take the situation ten years later—1957-1959.

YEAR	ALL COMMODITIES	FARM PRODUCTS	PROCESSED FOODS
1957	117.6	90.9	105.6
1958	119.2	94.9	110.9
1959	119.5	89.1	107.0
Total	356.3	274.9	323.5
Average	118.7	91.8	107.8

All this would suggest that the national statistics are being rigged? It is really an education to compare, say, the 1962 *Economic Report of the President* to later reports for the same years. In August 1965 statistical workers made a lot of changes. In one maneuver, interest was transposed from one side of the income equation to the other, thus raising corporate profits from $57.0 billion in the old report to $64.5 billion in the new. The corporations never got these profits and never paid taxes on them, yet they are in the statistics. How can the 1977 *Economic Report* show "100" for the 1967 period even though farmers drove tractors in protest in 1977 because of low prices. In the 1966 *Economic Report*, data are presented this way.

YEAR	ALL COMMODITIES	FARM PRODUCTS	PROCESSED FOODS
1957	99.0	99.2	97.9
1958	100.4	103.6	102.9
1959	100.6	97.2	99.2
Total	300.0	300.0	300.0

It averages out at 100 as if by magic. Now if you look at the 1977 *Report*, you'll see the all commodities index at 88.6 points for 1957, not 99.0 points as it appears in 1966 *Report* data for the year 1957. Agriculture hasn't had a balanced base period since 1946-1950, and computations have to be made from a balanced base to be valid.

The next argument posed by the people in government is that parity for agriculture costs too much. Actually, parity doesn't cost at all in an economic sense. *It supports the national income on an earned basis.* We're paying more to the interest mill than we're paying to the farmer. It is the interest mill that is feeding inflation into everything, even into cans of food on the shelf that were manufactured years ago. Officials can manufacture national income by inflating the money supply, but these are watered dollars and borrowed dollars. They get bottled up in the capital pool and are not drawn out to be circulated on an earned basis. When used, there is interest—and interest compounds itself chain letter style. This is the reason that bank liquidity—which is where 66% of the operating money is kept—goes down as farm income goes down. Today there is little bank liquidity. The banks are in fact loaning out the float. The only time bank liquidity jumped a little was when the Russian grain deal in 1972 triggered parity prices for most of agriculture for a few months. Once this temporary farm prosperity passed, bank liquidity started down again. If common citizens understood what parity prices meant to the economy, and not just to farmers, they'd be out at the farm gate with shotguns and prevent the farmer from letting his production go to market at less than parity.

One may ask—if the figures are rigged, how can you use government data to compute honest parity? You really can't. You have to come at it another way using your own data. Up to a few years ago, there were indexes that had not been rigged. *Wall Street Journal* carried a Dow-Jones Futures Commodity Index and a daily Dow-Jones Spot Commodity Index based on 1924-1926 equals "100," a fairly decent index. This umbilical to a stable period has now been cut. Reuters of the United

Kingdom has traditionally used 1931 as "100." The *Associated Press* index has been 1926 equals "100." You have to know what the base period is to understand what it means. Now there are commodity indexes. The problem is that with all this rigging, we no longer have a good *all commodities index*, or a good *processed foods index*. About all one can do is take figures from the *Economic Report of the President* and strike a new multiplier based on differentials in the rigged figures. It would be preferable, of course, to structure a new index based on actual *all commodity* prices and all processed food prices, with a few other composites thrown in. The late Arnold Paulson of National Organization for Raw Materials made some of these computations. Remember, he didn't have the computer time or the resources to handle this the way it might be handled by a dedicated college today. Yet he came up with parity figures based on a 1955 update and a 1910-1914 equals "100" that were quite close.

		End of 1976 (using 1955 update)	November 1977 (1910-1914 = "100")	End of December 1990
Wheat	(per bushel)	$7.01	$6.96	10.30
Cattle	(cwt. all grades)	62.00	59.41	110.20
	(cwt. choice)	77.00		
	(fat cattle avg.)		74.41	87.98
Rice	(cwt.)	15.18	15.47	22.91
Milk	(cwt. all grades)	13.06	12.95	19.95
Soybeans	(bushel)	8.10	8.19	12.129
Milo	(per pound)	.713	.3515	.52057
Hogs	(cwt. all grades)	62.00	62.50	92.50
Barleys	(bushel)	3.73	3.90	5.775
Corn	(bushel)	5.43	5.05	7.479

These were Paulson's computations. Our update is in the last column. A quick look at commodity prices reveals how the situation has worsened. Give or take a few cents, these were relative honest parity figures. Now look at your *Wall Street Journal*, June 1, 1982—wheat in Kansas City, at $4.0675, corn in Chicago, $2.585 and so on. You don't have to be a figure wizard to know that $4.0675 wheat divided by $10.30 (1910-1914 equals "100") is 39%. Just about all farm crops, some specialties excepted, were less than at half parity then, and they are worse now. And that's the reason the countryside is being emptied, and why it is taking over 25% of the national income just to mount relief agencies and so-called social programs. This explains why Reagan could create billions of new money at a whack and be called a conservative.

The common sense public thinks it is to its advantage to have low food prices. At least eight administrations have agreed, and for this reason the nation has maintained a cheap food policy. Is it really in the self-interest of consumers to have cheap food? The answer is *No*. The housewife pays a terrible tariff for the fiction of low food prices. Let's illustrate this. The family spent about twenty-five cents of its dollar to feed the family in the 1946-1950 era. At that times taxes took about twenty-five cents of the national income dollar. Today it takes about sixteen-plus cents to feed the family. But because the farmer has not been getting paid, the exchange equation has faltered. This has prompted the government to cover up by structuring all sorts of agencies to employ the people, all sorts of relief programs. Because the farmer was short-changed, it took debt and war spending and fear of communist aggression and foreign give-aways to sustain the prosperity. As a consequence it now takes forty-four to forty-five cents of the dollar to feed the tax collector. (With Social Security added, the tax lug is approximately fifty-five cents of the national income dollar.) Now sixteen cents for food and forty-five cents for the tax collector adds up to sixty-one cents, not the fifty cents when food cost a bit more. If there isn't proper division of income between different sectors, then the sectors can't consume each others' production. Generally, when there is farm parity, there are surpluses in the treasuries of taxing units, and there is liquidity in the banks.

We're told that if farmers get a parity price it means inflation—and it may mean higher prices temporarily as the economy adjusts itself. Farmers got parity prices briefly in the 1973-1974 era because of the Russian grain deal. This meant about 10% inflation, but remember there were other factors. There was a well choreographed oil crisis. The Fed kept on pumping money into the economy the same as if farm prices had not climbed. One thing is certain. As debt creation continues, public and private debt will double at least each decade, as it has since 1950. This has meant a public and private debt of $4 trillion by 1980, close to $12 trillion at 1990.

Parity for agriculture and other raw materials is the only non-inflationary course to take. There are only three ways to bring money into circulation. You can directly issue the money via government as provided for in the Constitution. This leads to inflation. You can tax money from people and circulate it, or use the credit mechanism, such as we do through the Federal Reserve. This leads to bankruptcy because it dissipates the savings and future earnings of our people. The only sound way to bring money into circulation is through the sale and production of raw materials. These dollars draw on the capital pool, are earned, and return to the capital pool with each cycle. The sale of 1,000 bushels of corn at, say, $2.50, draws on the capital structure to the tune of $2,500. If corn is $1.00, only $1,000 is drawn and earned into circulation. When the producer spends the dollar received for his product from nature, he passes the purchasing power to the next man and the next. The units of raw materials—new wealth—are transformed by industry into other forms of wealth and become permanent assets of the society. This isn't true of either the dollar of issue or the credit dollar. Since the gross farm income dollar represents something like 70% of all new wealth income and largely gauges the industrial demand for other products of new wealth, the relationship of agricultural income to national income on an earned basis becomes the governing factor of the economy.

Do not the economic advisers know this? Do not the bankers know it? Unless we take the position that these people are stupid, we have to assume they do. It is probably the most important discovery since the signing of the Constitution. It is the key to the economic democracy. It is also the key to undoing economic democracy. Knowing this relationship makes it possible for a few men who control speculative markets of the world to lower raw material prices to a world level. This forces the nation to exist by borrowing from savings and placing a mortgage on future income. It forces the inflationary issue of money. Since the American public does not comprehend this fact, they go along with the idea that an economy can use agriculture for a punching bag while compounding debt. Government programs invoked to plug the gap merely hand off inflation on a chain letter basis.

Parity is much more than a price comparison. It is a measuring unit that distills into one index figure production costs, the state of the arts in terms of inventions and technology, economic well-being, and income. It assumes and absorbs into its composites the historical reality that technology will not likely benefit one sector of an economy more than another. It is true, parity as a concept has limitations. It is not easily comprehended by economists who often deal with it.

Wouldn't parity hurt exports and therefore America's balance of payments? Under conditions of full parity, it would require possibly 50% less exports to return the same dollars in terms of payment balance. Cheap and free exports do very little to repair the international exchange imbalance. The import-export equation will be covered in another lesson.

Why do some businessmen fight farm parity? Many businessmen believe that all gain is not really earned, but achieved at the expense of others, and they therefore believe that in denying parity to agriculture greater gain will be achieved as business profits. They forget that business principles are not economic principles—and that economic principles requiring par exchange ultimately govern over business principles one way or another. Par exchange is a natural economic requirement. The only way to conquer a natural law is to obey it.

If grain prices are at parity, won't this hurt the cattle rancher or feedlot farmer? No. Parity for the six or seven basic crops that constitute 75% of the harvested acres has a self-adjusting effect on the rest of agriculture without regimentation of men, capital or production resources. Cattlemen get in a bind not because grain prices go up, but because they go down. Demand dries up long before the market is saturated.

Cattlemen have taken their whipping at regular intervals, the situation often being worsened by import invasion. Red meats, processed and boned, are imported year after year, the net result being displacement of feed grains, each pound of meat displacing seven pounds of grain on the average. Much of this meat is produced on $10 an acre land.

We estimate that without import invasion, each million one-thousand slaughter animals would require an additional cow herd of approximately 1,166,666 head plus an additional 33,000 bulls. It would mean the need for an additional 333,000 mother cows to produce those animals. The cattleman cannot be hurt by parity grain prices. He can only be hurt by lack of farm parity because a lack of parity authors poverty and makes it impossible for the population to eat properly. One key to low farm prices has been dislocation between farm commodities. Such dislocations disappear promptly once the basic crops get to and hold at parity.

Won't parity prices for agriculture create unmanageable surpluses? Again, no. This is not possible when farm raw materials are priced at parity because full parity prices set up the credits with which the population can consume the production. The only time there aren't surpluses is when prices are at parity for farm raw materials. The big surpluses have all been piled up when prices were at less than parity. At less than parity the income needed to consume the production is not created. Supply without effective demand cannot be consumed.

The farmer can get parity only with the aid of Congress. Congress should pass a law that mandates payment of parity prices as raw commodities enter trade channels. This is not unlike a minimum wage law. It simply would keep the trades from stealing farm production at the tailgate. It would not be necessary to regulate all crops. A computed parity on six or seven storable commodities that account for 70 to 75% of the harvested acres would do the job. Crops that spoil quickly—vegetables, certain fruits—could not be used in any formula. Such a base would permit the play of the market to function for minor commodities, keeping them very close to parity. Parity must be maintained at the market.

The loan program concept embodies a fundamental error. It permits buying power to be brought into existence and farmers to spend that money without the goods actually moving through the system. When the so-called stored surplus is eventually sold into trade channels, the farmer has already been paid for it, and he has spent the money. All he now gets in payment for the surplus in a loan program when it is sold is the price difference between the loan received and the market price at that time. Under a sound parity program, the farmer must be paid at full parity for production only when it enters the marketplace, and not one minute earlier. The loan system is a fundamental error because the farmer gets a loan price but in the end gives it away through inflation, or he creates stored inventories (because of distorted and shorted buying power) that hang over the economy and ultimately destroy the parity program. It can be fairly stated that government programs have been used to insure cheap farm prices and have forced perpetual expansion of debt, and these programs have at the same time assured built-in inflation. Parity can apply only to commodities that can be used. This is why the price must come from the marketplace at the time of the sale, not from loans or from subsidy payments. Only by passing through the market at full parity can farm crops generate a multiplier effect. A government that can set air fares, rail rates and prices on a pint of whiskey can also name the price at which basic raw farm commodities can enter trade channels. As for the subsidy concept, it represents little more than frustration economics—the tawdry business of keeping farm prices at a world level and giving a relief check to farmers to pace the rate of farm bankruptcy. The idea of the subsidy is to throw the farmer a bone so he won't get too restless for government comfort.

Parenthetically, it must be noted that the present type programs all were brought into being under the theory that it would be better to give the farmer a relief check to keep him satisfied politically than to permit him a proper price at the market. Farmers who understand economics are not asking for loans or supports. They are asking for parity at the marketplace. It is realized that to pay for exports of manufactured goods, financiers import many products from their acquired sources of production in other lands. These imports displace our own farm crops enough to create a spot surplus. Yet farm prices could be maintained by maintaining a price balance between six or seven of our basic crops—corn, wheat, barley, rye, soybeans, cotton, the foundation of all farm production—and when the general commodity index, which reflects industrial prices, moves 10% above a base period, then farm prices must also be adjusted to that index base. The big bugaboo about parity is "surplus," and the argument that agriculture will produce in excess of the economy's requirement. A good merchant maintains a price on his products. If he develops a surplus he has a sale and then spreads the markdown over the whole. He doesn't let the small overage set the price for the entire inventory. Even so, there is little evidence that a surplus could develop under parity conditions. Surplus pileups always develop when farm production is priced at less than parity. Parity for agriculture follows the other economic indexes. After an initial adjustment, full parity has a stabilizing effect because it takes bites out of the real cause of inflation—debts pretending to be earnings, and interest compounding debt chain letter style. Parity has been scuttled during recent decades largely as homage to free international trade. Yet exports below production costs really subsidize foreign buyers. Exports going into the Common Market often deliver more in taxes to foreign countries than the American farmers get as a price, and then those taxes are used by foreign governments to subsidize import invasion into the United States to further debase American farm prices. Opposition to full parity is based largely on conjectural economics and perceived self interest. The only ones who profit from less-than-parity prices are the great lending institutions that profit by keeping the economy trapped in ever-expanding debt.

Admittedly, government economists all say that parity won't work. Perhaps it would be more to the point to talk about track record. Government economists have had their way ever since the 1950s. As a consequence there hasn't been a balanced budget since the Truman years, when indeed farm raw materials last enjoyed parity for any period of time. If these gentlemen are so correct, then why is the American economy system poised at the brink of hyper-inflation and/or collapse? What we're talking about here has worked whenever it has been tried. What they're talking about has never worked, and it has been tried countless times throughout history. It seems that these economists who object to full parity for agriculture are letting failure go to their heads. It should stand to reason that nothing—not humming birds, bacteria or money—can compound chain letter style ad infinitum. At some point in time there will be a clash between mathematical ambition and physical possibility. If the present policy towards raw materials is continued, we will have between $25 and $30 trillion public and private debt by the year 2000, assuming the year 2000 chain letter game can last that long. Inflation has to follow public and private debt expansion as a direct ratio. Historically debt is the mechanism used to take the wealth of a nation from the many and put it into the hands of a few. It is not likely that this present economic policy can be followed much longer without having it preside over fantastic adjustments in political and institutional arrangements in this country. To put it bluntly, this means the nation will go into receivership. In political language, receivership is simply dictatorship, and when it happens it will come because economists have forty years of failure under the belt.

The economy will slide into a depression in any case. This cannot be stopped. It will wash out a lot of debt and set the stage for recovery. But there can be no real recovery without parity. That's the lesson history has taught. In fact, 1980 was probably the point of no return because the debt structure at that point could no longer be adjusted to a sustainable basis.

Future historians will no doubt bring this data into focus.

What would be the effect of farm parity prices for agriculture on industrial imports?

Parity for agriculture would secure parity for labor by bringing into focus the problem of import invasion. As it stands now, cheap goods flow to the high markets of the world. This import traffic has the same effect as importing cheap labor, and disemploys the American labor force. Thus the requirements of a parity equation for agriculture also make mandatory tariffs sufficient to make imports enter U.S. trade channels on par with goods produced at the American wage scale. Parity and structural balance cannot pertain only to agriculture. They must govern all sectors of the American economy, labor included, if full employment and a secure food lifeline are to be maintained. The impact of import invasion and resultant unemployment on the American labor force is substantial. Massachusetts Institute of Technology has estimated that approximately 25,000 jobs are eliminated for each $1 billion of direct private U.S. foreign investment. This defines how many jobs have been handed over to low cost employees the world over, chiefly in Asia. Workers thus unemployed in the United States pay out in more than lost jobs. A Johns Hopkins professor has computed that a 1% increase in unemployment rates increases 36,887 more deaths, 4,227 first admissions to mental hospitals and 3,340 commitments to prisons. In 1989 U.S. firms increased their investment abroad by $8.436 billion (to a new total of $31.765 billion). Wages in some of these countries are fifty cents an hour, even less. Low wages and even lower farm prices have resulted in out-migration from farms in these countries. As a consequence, there has been a demand for farm crop exports from the U.S. *at low prices* to accommodate the wishes of the multinational exploiters of labor and agriculture.

In order to stabilize the money and regulate the value thereof, Congress must achieve structural balance or parity. The two are synonymous. Parity is not just something for agriculture. There has to be a parity for labor, for business, for interest. This is what is meant by structural balance. For all practical purposes, an economy based on structural balance is one based on par exchange. Unfortunately the economic managers now refuse to evaluate this requirement of the exchange economy. For the time being, the interest mechanism is being used to fly another mini-cycle across the economic landscape. This distortion of logic has invaded every area of society. If par exchange is not allowed to intervene, if structural balance is not achieved as a consequence of parity, then the new era of history will be quite different from the democracy Americans have known. The point here is that parity and only parity can answer the requirements of the Constitution—"to coin the money and regulate the value thereof."

Because farm production at parity governs earned national income, its maintenance can be said to support national income. That will be the subject of Lesson 9.

LESSON 9

The near total eclipse of the family farm as a mechanism for broad-spectrum distribution of land, and therefore broad-spectrum distribution of earned income, is generally justified on grounds occupied by international traders. The bleatings of one business class for free international trade as a remedy for mal-distribution of earned income ought to be colored by its "self service" status, but it isn't. And the fact that free international trade is—in the words of John Maynard Keynes—"a desperate expedient to maintain employment at home by forcing sales on foreign markets and restricting purchases," a policy "which, if successful will merely shift the problem of unemployment to the neighbor which is worsted in the struggle"—and is thus the "predominant" cause of war. We have to admit that raw materials economics has to watch in wonderment this "desperate expedient" and the obstinate refusal to accept what is apparent to anyone possessed of uncommon good sense. The classical economist beat their chests as they utter Tarzanesque cries against war, and the political liberals moan their sorrow to oil over the fact of poverty, and yet the cause of both is rejected the way Fido rejects a mock hamburger. Schoolmen sidestep the issue by footnoting the underconsumptionists, and relegating them to intellectual underworld status. And the great fabricators, the great grain companies, and the kept firms of the military-industrial-university complex, all translate their demand for "more" into public policy.

But the fact is that a shortfall in domestic income cannot be translated into other than war by an access to foreign markets. Clarence Ayres argued with great persuasion that "all the wars of modern times have been economic in origin . . . a struggle for markets," a result of "deficiency of consumer purchasing power."

Thus the consequence of sinking parity and the failure of the multiplier to first deliver adequate national income, second to distribute the same. Admittedly, export traffic promises relief from "over-production," it being understood that the foreigner must first harvest raw materials or fabricate a product in order to create the credits needed to buy. Failure to create such credits has not undercut the "profitable truth" on which export trade relies, namely credit. Obviously Third World and less industrialized nations simply cannot buy unless semantic magic translates *selling* into *gifting*. Thus, in the end, commodities are literally given away at the expense of bond purchases or American taxpayers (since the government is often a guaranteeing agency).

Default is a fantastic blessing to business institutions when foreign sales are guaranteed by the government. The rule of "72" says that the rate of interest divided into "72" defines the number of years it takes for a debt to double. At 10% interest a debt doubles in 7.2 years. This means that when a debt is kept alive a little over twenty-two years, the interest collected has returned the amount of the debt three times, and yet the debt is still owed. If there is a default, the lending agency can collect just one more time under the provision of a government guarantee.

The cycles of war correlate perfectly with the cycles of debt, and the public policy that permits reciprocal imports to undercut domestic raw materials producers, factories, even services.

When John F. Kennedy was a Senator, he hinted at the

TABLE 5
COMPARISON OF NATIONAL INCOME
TO ANNUAL IMPORTS OF ALL GOODS

YEAR	NATIONAL INCOME (BILLIONS)	IMPORTS (BILLIONS)	PERCENTAGE OF NATIONAL INCOME FOR IMPORTS
1929	$87.8	$4.3	5%
1930	75.7	3.0	4%
1931	59.7	2.0	3.5%
1932	42.5	1.3	3.1%
1933	40.2	1.4	3.6%
1934	48.9	1.6	3.3%
1935	57.0	2.0	3.5%
1936	64.9	2.4	3.7%
1937	73.6	3.0	4.1%
1938	67.8	1.9	2.9%
1939	72.7	2.3	3.1%
1940	81.6	2.6	3.2%
1941	104.7	3.3	3.2%
1942	137.7	2.7	2.0%
1943	170.3	3.3	1.9%
1944	182.6	3.9	2.1%
1945	181.2	4.1	2.2%
1946	179.6	4.9	2.7%
1947	197.2	5.7	2.9%
1948	221.6	7.1	3.2%
1949	216.	26.6	3.5%
1950	240.0	8.8	3.7%
1951	277.0	10.9	3.9%
1952	290.9	10.7	3.7%
1953	305.0	10.8	3.5%
1954	298.3	10.2	3.4%
1955	324.0	11.3	3.5%
1956	349.4	12.6	3.6%
1957	364.0	13.0	3.6%
1958	367.7	12.8	3.5%
1959	399.6	15.2	3.8%
1960	412.0	14.7	3.6%
1961	427.3	14.7	3.4%
1962	457.7	16.2	3.5%
1963	481.1	17.0	3.5%
1964	514.4	18.6	3.6%
1965	566.0	21.4	3.8%
1966	620.6	25.4	4.1%

TABLE 5
(continued)

YEAR	NATIONAL INCOME (BILLIONS)	IMPORTS (BILLIONS)	PERCENTAGE OF NATIONAL INCOME FOR IMPORTS
1967	654.0	26.7	4.1%
1968	714.4	33.0	4.6%
1969	763.7	35.9	4.7%
1970	798.4	40.0	5.0%
1971	859.4	45.5	5.3%
1972	941.8	55.6	5.9%
1973	1,064.0	69.5	6.5%
1974	1,136.0	101.0	8.9%
1975	1,215.0	96.9	7.8%
1976	1,359.8	120.7	8.9%
1977	1,525.8	147.7	9.7%
1978	1,724.3	172.0	10.0%
1979	1,924.8	206.3	10.7%
1980	2,117.1	244.0	11.5%
1981	2,352.5	259.0	11.0%
1982	2,518.4	242.3	9.6%
1983	2,719.5	256.7	9.4%
1984	3,028.6	323.0	10.7%
1985	3,234.0	343.6	10.6%
1986	3,437.1	368.7	10.7%
1987	3,678.7	402.1	10.9%
*1988	3,984.9	441.0	11.0%
*1989	4,233.3	473.2	11.2%
*1990	4,418.4	495.3	11.2%

Source of Data—U.S. Department of Commerce.
*Source of estimate—Economic Indicators.

greatest delusion of our times. The committees he served had taken a look at the consequences of cheap imports, chiefly cotton. He gave a capsule report to a group of textile manufacturers. "Often you will find that while the import item may represent only 1% of a given commodity in proportion to the domestic production, the fact is that this 1% quoted at lower prices can seriously disrupt an entire industry."

This, of course, is exactly what has happened to American agriculture.

As we have seen, there have been no surpluses in terms of population growth—except seasonal—in the United States since well before WWII. Yet cheap imports—perhaps only a single percent in terms of domestic production—have served to undercut prices for American producers, driving the income generating capacity of the commodity producing machine well below parity. Faltering parity short circuits buying power and sets up the apparition of inventory buildup, a crowded pipeline—in a word, surplus.

We do not have room to graph each year for each commodity back to 1910-1914. But we have charted the shortfall in food production in terms of composite figures for each of the years illustrated. In other words we seem to produce between 4 and 6% less each year than it takes to feed the nation. This means we import more than the 1% John Kennedy said seriously disrupted an industry. Similar charges could be constructed for automobiles, for gas and oil, for home heating fuel and other necessities, generally with the finding that there is a shortfall between consumption and production in the U.S. Statistically, this is reflected in the negative trade and payment figures the Department of Commerce publishes monthly.

To feed the nation, we import the 4 to 6% shortfall in production, and this means we plug in prices based on eight to ten cents an hour labor in many cases. The process keeps American farmers producing at a composite called "world prices," which are half high enough to make an average well managed farm cash flow.

Even more important, these world prices undercut the

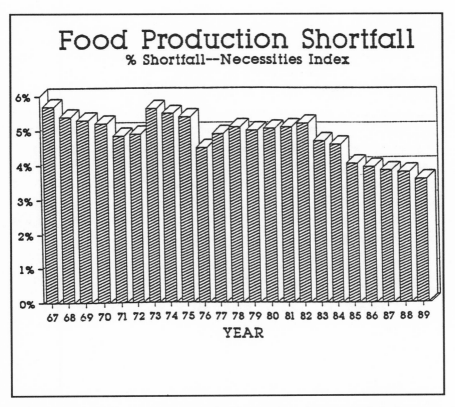

Food Production Shortfall
% Shortfall--Necessities Index

YEAR

generation of earned income both at the raw materials harvest level, and at the national income level. Under these circumstances, consumption can be kept alive only with new debt, social programs and—when all else fails—war!

The book *Unforgiven* carries a subtitle, *The Story of How America has Exchanged Parity Agriculture for Parity War*. In fact that simple whiplash line is an adequate precís for the content of the full book. Everything else is elaboration. The eclipse of parity agriculture has forced the public works program of war or preparation for war. War material is fabricated at par with wages and capital costs. Production employs a vast workforce and consumes raw materials in staggering quantities, also at parity, generally also on the basis of debt. Only the food component of ritualized killing is held to half-parity prices. Unfortunately consumption via destruction is suicidal. Such a public policy is bound to result in a price inflation.

The mechanism for making good nature's laws of physics can be discovered easily in the economic data assembled by the government, if we have the wit to look. These data tell us that when exports as a percent of national income exceed 6.5%, at that point the process presides over outmigration of companies and jobs to foreign climes and mounting unemployment, underemployment, homelessness and social programs to deal with the same at home. When imports as a percent of national income exceed 3.5%, the same apparitions decorate the economic landscape.

Thus one has to read with abject disbelief this passage in the 1991 *Economic Report of the President . . .*

> • The United States and other nations are endeavoring to strengthen, extend, and modernize GATT's rules governing international trade, as well as to reduce trade barriers worldwide.
> • Long-run U.S. goals of multilateral trade liberalization are embodied in the positions taken by the United States in the Uruguay Round. These include extending GATT discipline to trade in agriculture, textiles, services, and intellectual property; ensuring that developing countries take on the full obligations of GATT; establishing explicit international rules for foreign investment; and making the GATT dispute settlement mechanism swift, fair, and effective.
> • Successful completion of the Uruguay Round is important to the future growth and prosperity of the United States and the world.
> • The primary thrust of U.S. trade policy is to use multilateral discussions and for GATT and the Organization for Economic Cooperation and Development to promote free, rules-based trade. Indeed, the multilateral Uruguay Round negotiations are the President's top trade priority.

The federal agency now recognized by the initials G-A-T-T came into being in 1948, the year of the pilot measure to strike down parity, the Aiken bill. In time twenty-three nations joined this program for the purpose of impoverishing the world for the benefit of a few, namely lenders hard on the hunt for great debt potentials. The immediate objective of GATT was to remove the buffers against instant economic relief by annihilating all of the steps individual nations might take to protect their own from the ravages of trade flow. This trade flow is based on the lowest common denominator price available in trade channels. Usually it has meant six to ten cents an hour labor. Ultimately, by treaty, the several GATT rounds have hoped to take away from Congress the right to set farm policy.

The above climate of GATT seeks to overcome so that the free international trade "theory period" could enjoy its finest hour. This "finest hour" is being presented as a self-evident truth.

And that's the crux of the ideological debate over agricultural trade. The negotiators—preaching the sanctity of the market and the gospel of comparative advantage—maintain that each nation should get its food from the least expensive source, even if it means bankrupting its own producers and becoming more dependent on foreign suppliers.

The big international grain trading companies would be the big winners under the GATT system. They could operate for the first time with not even minimal concern for national borders, import quotas and tariffs, export restrictions or variations in international health and safety standards. They would be able to buy commodities cheap from countries that need the foreign exchange to cover their debts. With the elimination of GATT's Article XI, which currently allows countries to restrict exports, the big traders would be able to buy and export commodities even from nations with starving people. They'd sell the commodities at a profit to industrialized and oil rich countries which have the cash to pay. And if they sell at below the cost of production of the farmers in the importing nation, those farmers will be out of luck.

As U.S. commodity programs (officially termed "price stabilization" programs) are dismantled and the markets become more volatile, the commodity futures traders at the Chicago Board of Trade and the Chicago Mercantile Exchange would have greater opportunity to "buy low, sell high" and in general manipulate the markets even more than they have in the past. Global trading via satellite would soon command the world scene.

The conclusion that arrives, soon enough, would bowl over a

steer in an abbatoir if the figures were a sledge hammer. As exports as a percentage of gross national income move beyond that 6.5% cutoff point, we ship our buying power out of the country. Jobs are lost, and the trade turn of goods and services evaporates. Each time the perceived magic of exports gain an upper hand over parity, it always becomes a case of a few billion in export income annihilating consumer income billions four-fold, even more.

Carl Wilken put this in perspective with the following statement.

"A free trade mirage has no basis for existing . . . A 5% increase in domestic buying power is generally equal to all foreign trade. Our domestic price level, therefore, becomes the first order of business. Furthermore, our domestic population must consume those imports taken in payment for exports. As a result, both our domestic and foreign trade depend on the income of a domestic population."

If we import an automobile, American consumers first lose the income domestic production of an automobile would have created. Something other than the manufacturing function is displaced by the import. The next loss is the value of the imported automobile itself. If grain is shipped overseas, the U.S. loses the economic value of jobs usually created in handling and using that grain.

Data readily available in the *Economic Report of the President* reveal that with our present national income of, say, $5 trillion, we are exporting 8% of our national income and importing 10%—both well outside the parameters of 6.5% and 3.5% rule.

Having annihilated the very heart of free trade—if we accept inherent contradictions as a signal of defeat—the question surfaces as if to sour the landscape. It asks whether the world—especially countries without a full line of raw materials—could survive without free trade.

The answer is yes—if we have the wit to impose regulated equity of trade. The point here is that any nation is obliged to regulate what is shipped out. As it is, we ship out corn, lumber, hides, whatever, and import dressed meat, automobiles, leather goods—and corn, lumber, hides, etc.—always undercutting the American parity and putting American workers out of work.

Regulated equity of trade, simply stated, would require an American parity price for imports, and hand to the foreign supplier the right to collect for his product either American commodities or manufactured goods at an American price. Instead of driving foreign raw materials producers and manufacturers into poverty, improved earnings would enhance the world's standard of living. This, after all, was the plan of the Founding Fathers.

Contrary to common opinion, the intellectuals in America understood the curve of history. "Two systems are before the world," wrote Lincoln's economic adviser, Henry C. Carey, the son of Matthew Carey, a close collaborator to Ben Franklin and LaFayette. In *Harmony of Interests*, Carey went on to explain . . .

> Two systems are before the world; The one looks to increasing the proportion of persons and of capital engaged in trade and transportation, and therefore to diminishing the proportion engaged in producing commodities with which to trade, with *necessarily* diminished return to the labor of all; while the other looks to increasing the proportion engaged in the work of production, and diminishing that engaged in trade and transportation, with increased return to all, giving to the laborer good wages, and to the owner of capital good profits. One looks to increasing the quantity of raw materials to be exported, and diminishing the inducements to the import of men, thus impoverishing both farmer and planter by throwing on them the burden of freight; while the other looks to increasing the import of men, and diminishing the export of raw materials, thereby enriching both planter and farmer by relieving them from the payment of freight. One looks to giving the *products* of millions of acres of land and of labor of millions of men for the *services* of hundreds of thousands of distant men; the other to bringing the distant men to consume on the land the products of the land, exchanging day's labor for day's labor. One looks to compelling the farmers and planters of the Union to continue their contributions for the support of the fleets and the armies, the paupers, the nobles, and the sovereigns of Europe; the other to enabling ourselves to apply the same means to the moral and intellectual improvement of the sovereigns of America. One looks to the continuance of that *bastard* freedom of trade which denies the principle of protection, yet doles it out as revenue duties; the other to

extending the area of *legitimate* free trade by the establishment of perfect protection, followed by the annexation of individuals and communities, and ultimately by the abolition of custom-houses. One looks to exporting men to occupy desert tracts, the sovereignty of which is obtained by aid of diplomacy or war; the other to increasing the value of an immense extent of vacant land by importing men by millions for their occupation. One looks to the *centralization* of wealth and power in a great commercial city that shall rival the great cities of modern times, which have been and are being supported by aid of contributions which have exhausted every nation subjected to them; the other to *concentration*, by aid of which a market shall be made upon the land for the products of the land, and the farmer and planter be enriched. One looks to increasing the necessity for commerce; the other to increasing the power to maintain it. One looks to underworking the Hindu, and sinking the rest of the world to his level; the other to raising the standard of man throughout the world to our level. One looks to pauperism, ignorance, depopulation, and barbarism; the other to increasing wealth, comfort, intelligence, combination of action, and civilization. One looks towards universal war; the other towards peace. One is the English system; the other we may be proud to call the American system, for it is the only one ever devised the tendency of which was that of elevating while equalizing the condition of man throughout the world.

So much for the theory period in which we live. The Soviet Union is now emerging from a theory period called communism. It took nearly eight decades for the U.S.S.R. to produce statesmen who had the vision to ask, *What kind of damn fools are we?* And with that observation, the make-believe world of communism started to unravel.

We have still to find a leader who will look at our GATT-based free international trade and ask, *What kind of damn fools are we?* Foreign capital from Japan and Arab nations—where hatred of America has achieved an art form—is now buying up American institutional arrangements, and stands ready to make economic decisions for the U.S., all because of our child-like faith in theory period instruction.

LESSON 10

It is an article of faith among most economists that vast benefits are conferred on the community by the monetary system, thus its willingness to abide a money policy as dishonest as rigging the scales at a sales barn. Indeed, a monetary policy that creates money for one sector of the economy and permits imbalance with the consuming sector is a lot like having a weights and measures policy.

We have demonstrated the physical relationships between the value of raw materials input and national income. The speculative trades have contributed only discord to the norm of production times price equals income for the raw materials producer first, and for the national income on a ratio confirmed by the state of the arts. And we have hinted at the par relationship that must feed back to raw materials production if economic convulsion is to be avoided from time to time. Of the balanced base periods "100," we elect to use the most recent, 1946-1950 to make a point.

Data presented here have been taken from the *Economic Indicators*, which is published monthly by the Joint Economic Committee under the supervisor of the President's Council of Economic Advisors.

	1946-1950 Avg (Billions)	1990 (Billions)	Percent Increase
National Income	$211.0	$4420.1	1995%
Total Wages	136.5	3244.2	2377
Farm Operator	14.2	49.9	251
Small Business	21.4	352.6	1547
Rentals	7.2	5.9	(18)
Interest	4.5	466.6	10268
Corporate Profits	26.9	299.95	1015

To compute the percentage increase from 1946 to 1990,

4420.1	(1990 national income)
-211.0	(1946-1950 national income)
4209.1	(change in national income)

$$\frac{4209.1}{211.0} \times 100 = 1994.8341\%$$

or 1995% in a rounded number. Each component can be computed in the same way, as the column of figures under "percent increase" readily indicates. Thus it can be seen that while national income increased 1995%, farm income increased only 251%.

Gross farm income cannot do much to support the national income under these circumstances. The mineral and other raw materials component does a somewhat better job.

We realize that figures and arrayed data terrorize most readers, but we ask all to remember that what is being presented here is not much more complicated than the proposition that two and two are four. Take the comparison of the value of total raw materials production and gross savings. For the twenty-five year period 1929-1953—which included years of property, depression, war and recovery from war—gross savings computed to be 95% of the total value of raw materials. Considering the fact that values for minor items have been unavailable, this complied with the basic premise of raw materials

TABLE 6
GROSS SAVINGS

YEAR	RAW MATERIALS PRODUCTION (BILLIONS)	GROSS SAVINGS (BILLIONS)
1929	19.7	15.5
1930	16.2	11.2
1931	11.5	8.4
1932	8.8	2.8
1933	9.5	2.7
1934	11.8	5.6
1935	13.2	7.9
1936	15.2	11.1
1937	16.6	10.8
1938	14.4	8.9
1939	15.3	12.7
1940	16.5	16.0
1941	20.6	23.0
1942	27.2	41.8
1943	30.9	47.4
1944	32.5	57.0
1945	33.5	48.5
1946	37.8	28.7
1947	46.5	25.3
1948	50.3	36.4
1949	45.4	37.0
1950	47.5	42.0
1951	54.8	51.8
1952	54.1	54.7
1953	54.0	55.8

Note: Average Gross Savings for the twenty-five year period 1929-1953 were approximately 95% of total value of all raw materials. The drift since the introduction of excessive debt has thrown the normal relationship out of whack.

economics is remarkable.

When the subtle changes in an economy are not allowed to feed back data for parity construction at the raw materials level, savings falter. In earlier times this meant an observed depression.

Basic agriculture—the nation's largest industry, still primes the national economic pump and in effect hires all subsequent labor in the whole. The only exception is that of labor hired and paid by new capital investment in new buildings and equipment, which only a prosperous agriculture can sustain.

It was this last concept that Keynes and others seized upon during the 1930s, because capital expansion took on the color of earned income—the kind turned by raw material production—and satisfied those who created money at a profit. But debt that pretended to be earnings could not be sustained by anything less than a prosperous agriculture.

The compromise for our times has been debt. Debt, not raw materials, has been made the prime support for national income, and this debt is now doubling itself at a frightening rate, dragging inflation, social programs, and massive unemployment in its wake. The lack of savings in the U.S. economy is more shocking in its content than the apparition of figures suggests. Yet it is debt pretending to be earnings that is compounding itself and calling on the factors of frustration for an ever increased injection to manufacturers a single dollar of additional income.

Economists who fail to follow the logic inherent in the law of the lever and the primacy of raw materials point to data for years since 1953. They note with triumph that the ratio of raw materials to national income seems to have evaporated in the early 1950s.

The answer, of course, is debt. Debt has been constructed to make the case that raw materials income is not primary. Since we have introduced twenty-five year periods in this primer before, a similar device for making a point should prove valuable in examining the role of debt. All of the data used here are available in official U.S. government publications, chiefly in the

Economic Report of the President, Statistical Abstract of the United States, Survey of Current Business, and *Economic Indicators.*

For the period 1951-1975, a twenty-five year time frame, the U.S. enjoyed $14.6 trillion national income. Yet this income was not attained without being mortgaged. During the above period, the gross public and private debt increased from approximately $565 billion to $3,410.7 billion, an increase of $2,845.7 billion. In 1975 the total gross savings of the United States was $200.9 billion—this from a total national income of $1,209.5 billion ($1.209 trillion).

Divide the total gross savings of the United States, $200.9 billion, into the total national income of $1,209.5 billion, the result stands on the rooftop and shouts the warning that it takes $6.02 of national income to earn $1.00 of gross savings. Obviously debt cannot generate the savings it takes to pay off debt.

Here is how the figures fit hand in glove with history. Remember, in 1949 Congress struck down parity. It took until the arrival of the Eisenhower administration to implement this economic suicide. That is why our classic twenty-five year period came to a halt in 1953.

The scenario must be understood, otherwise our raw materials economics make no sense, and the new magicians indeed can conjure money out of thin air. The following paragraphs were published as pages in *Unforgiven* in 1971. Twenty years later they cannot be changed.

When the Korean War broke out, a few in Congress wanted to go the WWII route again. They wanted to monetize raw materials again, and keep a balanced [economy], but the move was blocked because traders sensed great profits ahead if free international trade could be maintained during a war.

Therefore it became the judgment of the economic advisers that raw material prices should be scaled down in homage to international trade, war or no war. Thus there was only one way to operate. More debt had to be created, $40 billion of it a year to fight the Korean War.

When the war ended, debt creation was scaled back to $30 billion, which immediately cancelled out $25 billion "unearned income" that debt expansion had accounted for. The 1954 recession followed.

Full employment could not be maintained without either parity raw materials or more debt expansion. To keep from feeling the full weight of unjustified debt expansion, raw materials were pressured downward.

In 1955, $72 billion of credit was injected into the economy. President Eisenhower was promising the nation a $500 billion national income. On the basis of $300 billion of national income, $72 billion debt creation represented twice as much money as the nation had a right to borrow based on profits and savings. Savings started running out in 1957, and debt expansion fell down to $35 billion, or half the rate of 1955. The 1958 depression followed.

By this time the mandate for debt expansion became clear to those who wanted to go the whole route. The small confederation of Cassandras supporting sound economic procedures failed to carry the day in the general's tent.

In 1959, $69 billion debt was created to shore up the faltering economy, whereas farm prices were pressured downward. But by 1959, the operating loss of the economy had increased to a point where $69 billion was barely enough to keep the momentum. A year later, 1960, the presidential campaign was fought against a background of "we've got to get the country moving again."

President Kennedy did that. He duplicated Ike's tactic simply because he was taking advice from the same school of economists. By year, Wilken calculated the public and private debt injection as follows:

1960	$	61 billion
1961		79 billion
1962		80 billion
1963		84 billion
President Johnson took over and—		
1964		94 billion
1965		104 billion
1966 (Wilken's estimate)		115 billion

With each shot of credit, raw material prices continued a downward trend. Rural America and finally the American democracy were being put on the line. Such as massive shift in the economic body could not be accomplished without a real sales job. Thus the string of policy papers, the *CED Report*, Michigan's Project 89 Report, the endless articles out of the land grant colleges, and finally the Food and Fiber Commission Report, all furnishing "laws" that demanded liquidation of the family farm.

There is no reason to run from these figures. But there is plenty of reason to reject the policies that author their existence. Ironically, it is the anatomy of the chain letter, not the sound physics of raw materials economics, that the nation's managers have embraced. It becomes difficult not to belabor this point.

It has been pointed out through the ages that nothing, not bacteria, humming birds, caged lemming or free roaming lions can double and re-double ad infinitum. Procreation must be offset by death within narrow limits. Much the same, the offspring of money in the form of interest must be matched by economic death. Some years ago *Acres U.S.A.* carried as front page fare the statistical tabulations of A. E. Hancock of Raymond, Alberta.

Money loaned at only 3% compounded annually will double in twenty-three years and 162.81 days, or in slightly less than 23.5 years. One dollar on interest from the birth of Christ to the present at only 3% interest would increase fantastically as shown by the following:

1.00	1	A.D.
2.00	23.5	
4.00	47	
8.00	70.5	
16.00	94	
32.00	117.5	
64.00	141	
128.00	164.5	
256.00	188	
512.00	211.5	
1,024.00	235	
2,048.00	258.5	
4,096.00	282	
8,192.00	305.5	
16,384.00	329	
32,768.00	352.5	
65,536.00	376	
131,072.00	399.5	
262,144.00	423	
524,288.00	446.5	
1,048,576.00	470	

2,097,152.00	493.5
4,194,304.00	517
8,388,608.00	540.5
16,777,216.00	564
33,554,432.00	587.5
67,108,864.00	611
134,217,728.00	634.5
268,435,456.00	658
536,870,912.00	681.5
1,073,741,824.00	705
2,147,483,648.00	728.5
4,294,967,296.00	752
8,589,934,592.00	775.5
17,179,869,184.00	799
34,359,738,368.00	822.5
68,719,476,736.00	846
137,438,953,472.00	869.5
274,877,906,944.00	893
549,755,813,888.00	916.5
1,099,511,627,776.00	940
2,199,023,255,552.00	963.5
4,398,046,511,104.00	987
8,796,093,022,208.00	1010.5
17,592,186,044,416.00	1034
35,184,372,088,832.00	1057.5
70,368,744,177,664.00	1081
140,737,488,355,328.00	1104.5
281,474,976,710,656.00	1128
562,949,953,421,312.00	1151.5
1,125,899,906,842,624.00	1175
2,251,799,813,685,248.00	1198.5
4,503,599,627,370,496.00	1222
9,007,199,254,740,992.00	1245.5
18,014,398,509,481,984.00	1269
36,028,797,018,963,968.00	1292.5
72,057,594,037,927,936.00	1316
144,115,188,075,855,872.00	1339.5
288,230,376,151,711,744.00	1363
576,460,752,303,423,488.00	1386.5
1,152,921,504,606,846,976.00	1410
2,305,843,009,213,693,952.00	1433.5
4,611,686,018,427,387,904.00	1457
9,223,372,036,854,775,808.00	1480.5
18,446,744,073,709,551,616.00	1504
36,893,488,147,419,103,232.00	1527.6
73,786,976,294,838,206,464.00	1551
147,573,952,589,676,412,928.00	1574.5

295,147,905,179,352,825,856.00	1598
590,295,810,358,705,651,712.00	1621.5
1,180,591,620,717,411,303,424.00	1645.5
2,361,183,241,434,822,606,848.00	1668.5
4,722,366,482,869,645,213,696.00	1692
9,444,732,965,739,290,427,392.00	1715.5
18,889,465,931,478,580,854,784.00	1739
37,778,931,862,957,161,709,568.00	1762.5
75,557,863,725,914,323,419,136.00	1786
151,115,727,451,828,646,838,272.00	1809.5
302,231,454,903,657,293,676,544.00	1833
604,462,909,807,314,587,353,088.00	1856.5
1,208,925,819,614,629,174,706,176.00	1880
2,417,851,639,229,258,349,412,352.00	1903.5
4,835,703,278,458,516,698,824,704.00	1927
9,671,406,556,917,033,397,649,408.00	1950.5
19,342,813,113,834,066,795,298,816.00	1974
38,685,626,227,668,133,590,597,632.00	1997.5

It is hard to visualize such an enormous sum. To spend such a staggering sum each of the earth's near four billion people would have to buy over three million automobiles each day—not for just a day or a year, but every day for over 1,000 years at today's prices. The system under which we live, wherein interest is paid on money created out of nothing, is the most ruthless bondmaster the world has ever known. Unless it is changed we will face complete loss of freedom under a world dictatorship. Strange to say, the United States has no system more rational than the chain letter economics illustrated above.

In the meantime, the rules of history prevail. A free society is being transformed into a controlled society. Within this context, the ratio of government spending to personal income, the rate of inflation and the accumulations of public and private debt are measures of the process of this conversion.

But this is getting ahead of the next lesson on foreign trade. In fact Lessons 10 and 11 should be read and studied as one. Go to the next page. We continue . . .

LESSON 11

"The greatest inventor the world has yet seen," said Thomas A. Edison, "was the fellow who invented compound interest." Indeed, this invention boggles the mind because it defies the laws of physics. Equally mind-boggling is the surplus fiction which hedges public policy. It was this idea of *surplus* that took national income out of the economic realm, and referred public policy back to a business equation, leaving macrophysics and macroeconomics standing in the wings. *Surplus* became a sacred fiction first, then a tradition because it best excused a head to toe recasting of the nation's institutional arrangements for doing business. *Surplus*—as a sacred fiction—made its debut during the Wilson administration, at which point pundits and copycats started yammering about farm inventory buildup in the public prints. We have noted before that between 1813 and 1914, a full century, American manufacturers remained sublimely indifferent to foreign markets. At the end of that era, a market basket of goods cost approximately the same as it had one hundred years earlier. Generally speaking, this era saw the act of production create the credits for consumption of production, the basic result—even with wars and depressions involved—being structural balance for the American economy, low unemployment and prices seldom ruptured by income invasion.

Using this platform, the U.S. built the world's greatest steel industry, and pushed almost all manufacturers to new heights. Indeed, the act of production created the money for consumption of production except when speculative raw materials "price discovery" got in the way.

This situation stuck like a sharp bone in the throats of the international lenders. Free international trade required a price dislocation between nations, thus the policy of the international lenders to "so reduce the purchasing power of the American people that they can no longer even approximately consume their own products."

The quoted words are from *The Breakdown of Money*, by Christopher Hollis. Hollis was a British historian on loan to Notre Dame University during the 1930s. He wrote,

> As long as that purchasing power was adequate, the American manufacturer was indifferent to foreign markets. But with domestic purchasing power reduced, foreign markets become essential to him. And, the more that he could be persuaded to look abroad for this market, the easier it would be to change his whole attitude toward wages. At present he is in favor of high tariff and high wages, for he looks on the working man as his customer. But, if he can be induced to look abroad for his markets, then wages become merely an item of costs and it is to the manufacturer's interest to reduce them as low as possible. If they are reduced—and the odium for reducing them of course, allowed to fall on the manufacturer—then American industry becomes at once a much more profitable investment for the financier, while the foreign goods can flow into free trade America to pay the interest on the foreign loans.

The opening gun in the drive to break the purchasing power of the American dollar had been fired as early as 1873 via the mechanism of silver, then an international money. From that date up to 1896, the price of silver broke from $1.32 an ounce to sixty-five cents an ounce. This was accomplished on the commodity exchanges by manipulation, much as the grain traders broke soybean prices on the Chicago Board of Trade in 1989. For every cent that silver went down between 1873 and

1896, the price of wheat in the high plains went down a penny a bushel also.

These probes continued well into the Wilson administration, at which time the international lenders struck pay dirt. Almost overnight the American republic amended its past and recast its future in terms of the price dislocation requirements of free international trade.

There was that short interlude—the 1940s—when a stabilization measure and the Steagall Amendment returned raw materials prices to parity, but this was all swept aside the minute Dwight D. Eisenhower took office. In homage to free international trade, farmers were cast into an economic abyss.

Harry S Truman called the 80th Congress a do-nothing Congress, the worst in history. It was this Congress that gave him an Aiken bill to sign, and then allowed the barest subsidies to keep American producers in business. Truman was sandbagged in this political game, threatened by a veto override, tired to death, and assured of his own defeat in the 1948 election by everyone who could write for print. Harry Truman had made his mistakes, but failing to understand the economics of the nation was not one of them. No one was ready to listen to statesmanship in 1948. The international flim-flam men won the day, and the plan the Council on Foreign Relations (CFR) types suckered USDA Secretary Charlie Brannan into accepting made the last years of the Truman administration look like a convention of free-trade fellow travelers, something Truman never was.

Suffice it to say that as long as American agriculture had parity, lenders of every stripe went hungry. At the end of the war parity years, the banks were chock full of money, but applicants for loans were few. In fact there was only a market for 14% of the available funds at one time. The minute USDA Secretary Ezra Taft Benson, who followed Brannan, struck down parity, farm operators started the painful process of consuming their capital, and loan applications soared. Government lending agencies were created or expanded to take risky customers off the hands of the commercial banks.

Unforgiven provides a few details that are now of maximum interest:

As prices continued to sink, one wave of pundits after another hit farmers with the productivity routine. The Committee for Economic Development, with Goldsmith's *A Study of Saving in the United States* in tow much of the time, did its hatchet work on a continuing basis, starting immediately after WWII. Policy papers such as *Toward a Realistic Farm Program, Economic Policy for American Agriculture* and *An Adaptive Program for Agriculture* explained to farmers how they had to get bigger, more efficient, and finally depart the scene. The Bible-style texts of Don Paarlberg and Dale Hathaway also arrived to help style public policy so that the job of liquidating agriculture might proceed without bringing on open revolt from the farmers, and by the time those volumes had been thumbed to death in the learning halls, Johns Hopkins Press struck off another entry, this one titled *Policy Directions for U.S. Agriculture.* The last was issued under the auspices of Resources for the Future, Inc., a non-profit organization based in Washington for the altruistic purpose of serving the powers that be. The writer, one Marion Clawson, came to the Resources for the Future Foundation via the customary route—out of USDA and the Interior Department. His message came to the public from the mature fortunes of the nation by way of the tax-supported instrumentalities that no longer knew which hand did the long range feeding.

Instead of support programs, Clawson fell back on the Brannan idea, direct payments to certain farmers, or institutionalized poverty, so those who remained could double their production, enjoy prices held down by imports, and starve to death in the end.

Pensioning farmers and speeding the exodus from the countryside could not solve either the farm problem or the city problem because rural America required the gross dollar, and without the gross dollar rural towns would starve as well. The sore of low farm income thus became the cancer that consumed fully a third of the American population by the end of the 1960s.

The "too many resources in agriculture" pitch failed to express or prove a concept of economic growth, because too many resources in agriculture could become too many resources in widget-making, entertaining or professoring, for that matter.

The mechanism for this debauchery of agriculture is not well understood by farmers even today, and because of this information shortfall, farm organizations talked nonsense and lost members from the beginning of the Eisenhower administration to the present, the process being exacerbated with the arrival of satellite technology.

Soon enough, prices became governed by an invisible currency in the skies linked to the monetary systems of almost every nation on earth, yet answerable to none, or to the requirement that without monetized raw materials there can be no profits or savings for the system.

> This electronic currency [wrote Harold Wills, a NORM associate] consists of a satellite rotating with Mother Earth so that it stays in place above Brussels, Belgium, and is linked to a banking network able to transfer the equivalent of cash between any two or more points on earth in a matter of seconds. An offer can be made, it can be accepted, and electronic entries can convert it into cash in currency of the selling nation as easily as when a farmer hauls a load of wheat to his local grain elevator and receives a check for his efforts.
>
> The farmer, of course, receives the price offered by the elevator and which may or may not have any relationship to parity. In global electronic transactions, the relationship is the same but the volume is much larger, a million or so bushels instead of a truckload. Also, there are more middlemen who must receive a piece of the action.

The American farmer has been told—often several times a day—how supply and demand function, and why "too many farmers" in effect annihilate farm income. And yet the figures and the charts all say these arguments are pap for USDAs trader clients who cherish an international price dislocation most of all.

It must be understood, of course, that surpluses are not generally the consequence of nature's vagaries. Except in cases of drought and natural calamities—such as 1816, when Mount Tambora on the island of Sumbawa, kicked twenty-five cubic miles of debris into the atmosphere (ten times more than Krakatoa) to circle the globe and cause the year without a summer—shortages are not created by weather, and surpluses are not the result of superhuman farmers creating bins and bushels of products beyond the physical capacity of a hungry planet to eat. Most of the time surpluses are the consequence of underconsumption made mandatory by prices either above or below parity. A parity price correctly computed off a proper base period "100" is the index for maximum consumption on a continuing basis.

Carl H. Wilken gave expression to both the problem and the solution in a paper styled *Pareconomy*, which has been preserved for us because he read the gist of it to a congressional committee.

Under our present system of operation at less than parity prices, we are fighting two surpluses—that of underconsumption and the natural surplus of seasonal production. Our attempts at production control have been rather futile because the program which we adopted was one of increased production. It is a well-known fact that when a farmer buys a rundown or overcropped farm he seeds it to clover, but no to produce less—to produce more.

Our soil building program, even though the farmers were led to believe that they were reducing production to obtain a price, will result in greater production. The same result would occur if we had a parity price relationship. The laws of nature would force the farmer to rotate his crops with legumes in order that he might produce more. In years of high prices we find that the acres in cultivation are reduced and that in years of low prices the acres of land in tillable crops are increased. But, strange as it may seem, the records indicate that we never have had a permanent surplus of farm products. Even with the tremendous underconsumption that prevailed in 1932, the record shows a net import of farm products in that year.

In July 1937 . . . the wheat market was in a strong position and our domestic grain buyers bought wheat for $1.14 and $1.15 per bushel as fast as it came to the market. But a few weeks later the men in control of the grain markets of the world forced the price of wheat downward about thirty cents per bushel. For every thirty cents taken off the price of a bushel of wheat they deprived the man on the factory payroll of thirty cents wages and they deprived the American people of $2.10 in collective income.

On the other hand, the formula means that whenever this nation will use its good common sense and stabilize farm prices in direct proportion to the national income needed for prosperity, our depression will be over. And it also means that the farm problem and the question of a price for raw materials is the number one problem of every farmer, laborer and businessman in the United States.

The 1935 *Yearbook of Agriculture* has statistics for corn, oats, wheat, barley, rye and flax for both 1928 and 1932. These data reveal that production for 1932 was a little over 1% less than in 1928. The pattern for pork, beef, mutton and veal was much the same.

The record shows further that in spite of the fact that over ten million people were unemployed in 1932, the U.S. consumed all but 148,000,000 pounds of this meat, or a little over one pound per capita. Yet we allowed that surplus to reduce the income of the United States an amount equal to the total value of all the farm products produced in the four-year period 1923-26, inclusive.

If the price for corn and other products had been kept the same in 1932 as in 1928, it would have been impossible to have a depression, unemployment, and hunger. In short, the U.S. had the wealth, but allowed someone else to fix the price of that wealth.

And yet the signal word for continuance of this frustration economics has been SURPLUS! The surplus code word was a fiction, of course. At the end of the eleven year price support era that saw four national budgets balanced—six years of war, five of peace—total stocks approximated $2.5 billion. The inventories for three basic storable crops were 256 million bushels of wheat, 2.8 million bushels bales of cotton, and 487 million bushels of corn. Such a corn supply was good for two months. The other *surpluses* constituted a bare inventory.

Data from Table 715, page 656, *Statistical Abstract of the United States*, 1944-1945, indicate that between 1933 and 1943—an eleven year period—U.S. exports of farm products totaled $8,723,787,000. Imports for the same period came to $12,786,725,000. The U.S. imported $4,062,938,000 more of farm production than was exported. In composite terms, the U.S. excess of imports of farm products over exports of the same products for the period amounted to about ten entire crops of wheat. Thus the surplus ploy was used to hold down farm prices before WWII. It was used to annihilate parity. Then it was used to shore up the intellectually indefensible premise on which the *CED Report* and the *Young Executives Report* was constructed. And this fiction is still being used by GATT to burn down the bridges so there can be no retreat to parity even if the present theory period ends and economic reality arrives.

Data covering the three decades between 1960 and 1990 suggest that, indeed, there have been no surpluses—except short

TABLE 7
USE GREATER THAN PRODUCTION
U.S. WHEAT AND COARSE GRAINS
(Million Metric Ton)

Year	Beginning	Production	Use	Imports	Ending	Percentage of Annual Production
60/61	105.6	178.8	166.7	0.6	118.3	93%
61/62	118.3	161.0	175.5	0.5	104.3	109%
62/63	104.3	159.3	170.8	0.3	93.2	107%
63/64	93.2	171.5	175.0	0.4	90.1	102%
64/65	90.1	157.5	172.9	0.4	76.5	109%
65/66	76.5	179.1	197.8	0.3	58.2	110%
66/67	58.2	180.7	189.7	0.3	49.5	104%
67/68	49.5	203.9	191.0	0.3	62.7	93%
68/69	62.7	197.6	188.9	0.3	71.8	95%
69/70	71.8	201.0	200.4	0.4	72.8	99%
70/71	72.8	182.9	201.6	0.4	54.6	110%
71/72	54.6	233.6	215.1	0.4	73.4	92%
72/73	73.4	224.1	250.0	0.5	48.0	111%
73/74	48.0	233.3	250.5	0.3	31.1	107%
74/75	31.1	199.4	203.7	0.6	27.3	102%
75/76	27.3	243.3	235.7	0.5	35.5	96%
76/77	35.5	252.8	228.4	0.4	60.3	90%
77/78	60.3	261.4	248.6	0.4	73.5	95%
78/79	73.5	270.5	272.7	0.3	71.6	100%
79/80	71.6	296.5	291.2	0.4	77.2	98%
80/81	77.2	263.1	279.1	0.3	61.6	106%
81/82	61.6	322.4	284.6	0.4	99.8	88%
82/83	99.8	326.0	287.7	0.6	138.7	88%
83/84	138.7	203.0	272.7	0.8	69.8	134%
84/85	69.8	307.6	294.8	0.9	83.5	95%
25 YEARS		5610.3	5645.1	11.0		100.6%

NOTE: *Coarse grains include corn, sorghum, barley, oats, rye.*

SOURCE: United States Department of Agriculture
                                       Foreign Agriculture Circular
                               World Grain Situation and Outlook
                                                February 12, 1985

AVERAGE ANNUAL PRODUCTION   *234.4 Million Tons*
AVERAGE ANNUAL TOTAL USE   *225.8 Million Tons*

term. And this stretches the general balance in farm production from 1910-1914 to the present, a period in which only short term surpluses have existed. And yet this fiction of *surpluses* has been used to literally kill off broad spectrum distribution of land and income for America, and thereby usher in fantastic changes in our political and institutional arrangements.

Data in Table 7, reproduced here, illustrate that ending grain stocks in 1985—at the height of the foreclosure mania—were less than stocks in the early 1960s. Ending stocks, as a percentage of annual use, actually decreased from 71% in 1961 to 28% in 1985.

Changed procedures in government figure gathering have made extension of Table 7 beyond 1985 confusing, if not impossible. Therefore these data for U.S. production and usage of corn and wheat are included for 1986 through 1990, 1989 and 1990 being represented by preliminary and estimated figures.

WHEAT
(Million Metric Tons)

YEAR	PRODUCTION	USAGE	STOCK CHANGE	AVERAGE PRICE
1986	2,092	2,197	-105	2.42
1987	2,107	2,689	-582	2.57
1988	1,891	2,397	-586	3.72
1989	2,037	2,225	-188	3.72
1990	2,739	2,328	+411	2.55/2.65

CORN
(Million Metric Tons)

YEAR	PRODUCTION	USAGE	STOCK CHANGE	AVERAGE PRICE
1986	8,250	7,410	+840	1.50
1987	7,072	7,699	-627	1.94
1988	4,929	7,260	-2,331	2,54
1989	7,525	8,113	-588	2.36
1990	7,933	8,020	-87	2.36

Using wheat and cotton as focal points, the great traders have constructed a Hitleresque "big lie" surplus syndrome. Perhaps there is more wheat than American bakers can roll into spaghetti in any year. Still, there has to be a pipeline. In an abstract form, the amount of wheat constantly in transit and storage—to iron out the fact that at harvest, there is a 365 day supply, and one day of demand—cannot be used. But it was used in 1972. The pipeline was emptied to supply the Soviet Union, and the USDA and Chicago Board of Trade made merry, declaring that between 50% and 75% of the crop went into foreign trade channels.

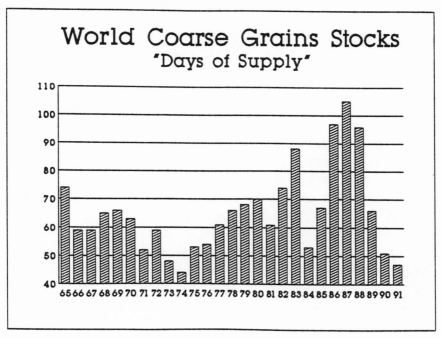

This graph, generated via computer, amply illustrates how surplus equals shortage. By ironing out the grains into composite figures, the common denominators reveal that in fact the pipeline is kept dangerously low in order to make the surplus fiction appear real.

LESSON 12

Now that we have arrived at the end of our primer course in raw materials economics, it is appropriate for the late Vince Rossiter to furnish a summary. Rossiter was the President of the Bank of Hartington, Hartington, Nebraska, and a confidant of both Carl Wilken and Arnold Paulson, as well as survivors of the old Raw Materials National Council. Later he served as president of National Organization for Raw Materials (NORM), during which time he wrote this lesson. This summary has been updated and edited for its present purpose, and yet it remains essentially as written before Rossiter's untimely death in 1990.

The National Organization for Raw Materials, [wrote Rossiter], more popularly known as NORM, has since its inception dealt with the need for economic equilibrium between the various sectors of the United States economy. What we advocate is a fair sharing of national income in its distribution between the various divisions of labor which will provide an equitable, reciprocal market for all of the goods and services produced, all of which can be paid for with the cash (earned income and profits) received in an annual economic cycle. The

5 or 6% of national income which goes to savings is all the credit that need be loaned back into the economy to provide a 100% market.

Early research acquired from a predecessor group known as the Raw Materials National Council, provided a base of economic statistics (see pages 53, 55, and 57) which supplied evidence of a remarkable relationship between the income derived from the sale of raw materials and total national income from 1929 through 1953.

The significance of this discovery cannot be overemphasized. It is not the product of conventional economics. It was not invented by economists. It is economic history created by practical human beings operating in a market which proved—whether raw material income was low or high—that national income was five times the value of "all" raw material income annually, and seven times the value of "only" farm raw material income.

Another study proved that farm raw material production, renewable annually, is approximately 70% of all raw material production in the United States under conditions of full employment and structural balance. The other 30% are minerals of various kinds, which are exhaustible over time. Thus farm raw material income is destined to become increasingly important as it becomes a larger share of total raw material production in the long term.

A third study developed the evidence that there is a long-term relationship between cash receipts from farm marketings and the payrolls of production workers.

Raw Materials National Council came to the conclusion that $1.00 of raw material income will result in the creation of $1.00 of factory wages and $5.00 of national income, again under conditions of normal debt use.

This is known as the raw materials multiplier effect on national income. It is the basic ingredient—not the only ingredient, but the basic ingredient—in maintaining economic equilibrium. Raw material prices should never be permitted to drop below a level of "100%" of parity with the average price level of all other goods and services. As the basic and initial monetary factor, it will maintain factory wages at a predetermined level, and in the process, create the income (cash plus savings) which will assure the "100%" consumption essential to a viable self-liquidating market process.

A prime example of the consistency with which the economy of the United States has functioned historically is the relationship between national income in any given year and the amount of money consumers had to spend on goods essential to life, such as food, shoes, clothing, beverages at almost exactly 70% of consumer expenditures every year from 1929 to 1955.

So, you say, why not continue on with these same statistical data from 1953 through 1982? Won't it prove, you ask, the findings of the Raw Material National Council conclusively?

Unfortunately the answer is, *No*, it doesn't.

NORM has always had a problem presenting what we considered a rather conclusive array of factual evidence on how the economy functions. We have concluded that it is the utter simplicity of the proof of our findings, prior to 1953, that was an affront to the "value free" approach of establishment economists who have a notion that there is an invisible hand directing our economy through the discipline of the so-called law of supply and demand in a mysterious natural order of things. Needless to say, non-economists are not permitted to move freely in this mystical arena, but fortunately they do and our work is evidence of this fact.

The Raw Materials National Council, [based on the exhibits revealed in this primer and numerous other studies] also realized that if it were true—as the evidence indicated—that $1.00 of raw materials income generated $1.00 of factor wages and $5.00 of national income, then the opposite of that equation would also be true. That is to say, every dollar that is lost to the economy due to underpricing of raw materials in the marketplace would have a REVERSE MULTIPLIER effect, which would deny the creation of $1.00 of factory wages and $5.00 of national income. [See the glossary entry, REVERSE MULTIPLIER, page 152.]

The evidence of this loss would be reflected in reduced employment, idle plant capacity, and an excess of consumers goods for which there is no buyer. All of these are rather familiar symptoms of economic malaise since 1953.

The degree of loss, which was not picked up by the injection of excessive debt, would be determined by the prevailing raw material price level—which held at an all time low of approximately 50% of parity during the 1980s. If the flow of excess credit is diminished, or if it is exhausted, then the parity ratio will be absolute as a market determinant for the production and sale of goods and services.

The apparent ability of the debt dollar to perform in the same 1-1-5 multiplier manner as a dollar of raw material income, when the economy is in near equilibrium, prompted a long period of irrational economic planning to greatly influence the U.S. economy earlier in this century.

There was a very influential school of economic thought which reasoned that debt is the answer to economic equilibrium and prosperity in the nation. This school declared that the multiplier of $1.00 of debt flowing through the economy could create enough velocity in the money supply to bring into being permanent savings. Needless to say, if this were the case the economy of the United States (and the world) would be rolling in prosperity right now—but this is not the case.

The economy of the United States and the world is walking a tight rope between inflation and depression because of this excessive public and private debt that was supposed to make the nation and the world prosperous. All it did was to provide a temporary respite from the depression that would have occurred much earlier as a consequence of steadily increasing losses of earned income and profits over the last thirty years. Now that there is no longer enough readily available credit to continue this Band-Aide debt injection, and there is no alternative solution, the economy flounders on the verge of chaos.

At this point one would expect a return to economic sanity and recognition of the historic economic and social facts of life that prevailed prior to 1953. At the same time, the constantly growing economic distortion in the sharing of national income, which has prevailed since 1953, has completely ruled out the relatively simple comparisons illustrated here. The inability to use similar comparisons with a degree of confirmation of the previous findings has been used to discredit the earlier findings of the Raw Materials National Council. Growing distortions of income share, coupled with resulting inflation of income totals because of a devalued dollar, prompted NORM to attempt to trace the negative multiplier impact due to underpayment for raw materials since 1953. It seemed to be a logical alternative. If the public was unwilling to believe the seemingly irrefutable evidence of the period prior to 1953, as clear as it is, then another approach became imperative. In order to do this, it was necessary to wait for nearly forty years. Not all of us can be sure that we have that long to wait for answers to perplexing problems. Many didn't make it and those that did aren't necessarily the ones who would do the best job of further analysis. However, the methods of analysis have not changed and the statistical material has improved and become more plentiful over time. As was the case prior to 1953, the information is clear and persuasive.

The reverse multiplier was put into place by the abandonment of full parity in 1953. It was decided, as a matter of government policy, "to let the farmers live off their fat for a while." The farm act of 1953 made secure flexible parity, with a prescribed range of 60 to 90% price supports. It came at the outset of the administrations of President Eisenhower and Secretary of Agriculture Ezra Taft Benson.

Despite the fact that the WWII stabilization measure and the Steagall Amendment, which provided 90% farm price supports from 1943 through 1952, had the effect of providing agriculture with an average of "100%" of parity prices for ten years through the economic stress of WWII, and the economic trauma of reconversion to a peace-time economy after 1946, it was abandoned in 1953. This set the stage for heavy injections of excessive debt, historic distortions of national income share, unprecedented federal budget deficits, record financial illiquidity, and a higher debt structure than has ever existed in all human history, all in hardly forty years.

Rather providentially for the United States and fortunately for NORM and its further research, the stabilization laws assured economic equilibrium for ten years during WWII and the post-war period up to 1953. This period provides a sound and relatively modern base from which to test the validity of the "reverse multiplier effect" of the negative 1-1-5 trade turn of raw material income.

The distortions, in the share of national income earned by each of the six designated sectors of national income, began to appear in 1952 because of the widespread anticipation of the move by Ezra Taft Benson, which eliminated fixed 90% of parity price support.

The Raw Material National Council had anticipated that this would happen and predicted that when it did, it would set the stage for economic chaos and ultimate depression. Ironically, a study copyrighted in 1949 had a specific reference to Social Security payments. It stated, "There can be no Social Security without price stability. Provisions for government doles become a myth as falling prices preclude taxation to pay bills."

The adverse effect of sub-parity farm prices became evident as early as 1960. Analysts began to discern that commercial bank loans were expanding at a rate of two percentage points more rapidly than the growth in bank deposits. When this was projected forward, it appeared that the commercial banking system would be as fully loaned and invested by as early as 1973 as it was on June 30, 1929, just four months before the beginning of the depression of the 1930s. And it did exceed "100%" on December 31, 1973. At the time, that was a very disconcerting bit of information to the bankers who were old enough to have a recollection of the depression of the 1930s. It was also bad news for those familiar with raw material economics and the negative trade turn attributed to the debt dollar. It was a valid and strong indication that the long run trend of the economy had turned down.

After sitting on the information for two years, in March of

1962 a group of bankers representing a major national banking organization, along with representatives of the Raw Materials National Council, decided to take the issue to Washington, D.C.

One primary concern centered around the rapidly declining share of national income that was being earned by the agricultural economy after 1952. The effects on the rural farm and business economy were adverse and growing. The decline in the ratio of loans and investments in relation to bank deposits affected not just rural banks, but it was evident in all of the banks in the United States.

The meeting was with Arthur Okum, then the newest appointee to the Council of Economic Advisers to the President of the United States. It was in his office that we were rudely awakened to reality with a definitive response to the question uppermost in our minds. Our question was, "What is the Council of Economic Advisers to the President planning for the agricultural economy of the United States in the near and more distant future?"

Okum was very open with us and readily admitted that his expertise was in other areas of economics. He said, "Frankly, I don't know much about agriculture, but let me call Dr. James Bonnen who is a specialist in this field."

Bonnen arrived a few minutes later and the discussion continues. We repeated our concerns for the future of agriculture and the latent effect it could have on the entire economy if nothing was done about farm prices. He responded with some of the same economic gibberish that we had heard, without exception, every place we had been.

Finally Okum interrupted the exchange between the committee members and Bonnen. He said, "Dr. Bonnen, it is getting close to lunch time. These people came to us with a specific question, and I believe that we owe them a specific answer. They are asking what the Council of Economic Advisers are planning for the agricultural economy in the near and more distant future. After what I've heard this morning, I too would like to know the answer to their question. What is our plan for agriculture?"

Bonnen paused for a moment and then with a gesture of up-turned empty hands, he said, "I'm sorry Dr. Okum, the Council of Economic Advisers has no plan for agriculture."

The realization that Okum admittedly knew little about agriculture, and that Bonnen, his specialist, admitted that "there is no plan for agriculture," at the level of the Council of Economic Advisers to the President, was very distressing to the committee.

The Washington attitude toward agriculture was both puzzling and frustrating to the committee. It can only be described as conviction that there were no problems in agriculture that would not correct themselves in the long run, and that the agriculture economy had a debt to society that was still unpaid. Our admonitions against further neglect of this sector of the economy, because of certain injury to other sectors of the economy, were of no concern to those we spoke to. They did not believe what we said.

The title of this lesson might well be, *There is No Plan for Agriculture.* However, more recent perception of events has almost convinced us that there was a plan after all. It might be better described as a "non-plan" rather than a plan. Or even better yet, it might be described as an "ono" plan; *oh no* you can't do this for agriculture and *oh no* you can't do that for agriculture. The way it has worked out for the last half century, except for the 1943-52 period when we had mandatory 90% price supports, we have always had the same "cheap food" policy that prevails today.

Harold Wills has told us that the only responsibility that any Secretary of Agriculture of the United States has ever had has been to "maintain the status quo" as the order of the day.

The "non-plan" for agriculture might be briefly outlined as follows:

• Permit the agricultural economy to drift with the economic tide as much as possible.

• Let it experience as much "natural" erosion of assets and resources as possible, short of creating a food shortage or higher prices.

- Permit, and possibly encourage, large non-farm corporations such as the largest insurance companies in the nation to acquire farm land and other assets in order to assure production if the family farm system collapses.
- Insist on further contributions by agriculture to the industrial manpower pool.
- Extract capital resources from agriculture to be invested in manufacturing industry.
- Require agriculture, as its responsibility to society, to provide a good market for manufactured goods.
- Demand of agriculture that all of these things "be done without an increase in total resources used and/or in the relative price of farm products."

If you are appalled by that litany of the provisions of the agricultural non-plan policy of the United States, join the crowd.

When the Congress of the United States "deregulated" farm prices, it denied agriculture an increase in "the relative price of farm products" at par with the average cost of other goods and service produced in the United States.

This act of Congress virtually forced American agriculture into the Russian mold of "perpetually cheap farm raw materials." But, the price of bread in the U.S. hasn't remained as cheap as it was less than forty years ago. Only the wheat that goes into bread has stayed cheap.

We cannot preserve the American private enterprise economy for all of the other sectors and exclude agriculture. We cannot sovietize agriculture with low deregulated prices on farm production, and at the same time, preserve democracy and personal freedom in the other sectors of the American economy. It just doesn't work that way and that is exactly what we have proved, once again, in the last forty years in the United States.

Establishment economics has been permitted to run amuck in the United States in the past forty year. It has misused and abused the American family farm enterprise. It goes back much farther than that, of course, to the early 19th century when David Ricardo promoted the so-called law of supply and demand. By its nature, agriculture must produce twelve months in advance of the requirements of society; and it has been the victim of supply and demand in every nation where the government either ignores or neglects its responsibility to protect farm prices.

I was pleased when John L. King, himself an economist, a graduate of Wharton's School of Economics, and also an instructor there, verified my opinion of economists or at least substantiated it to a large degree. King wrote, "It comes down to this: the establishment economics that are taught in the universities, proliferated in journals, regurgitated by councils of government, with all of its mountains of published output, has not advanced our capacity to control our economy beyond the late 1930s."

You may be interested to know that it is a distressing and a demeaning experience to be a banker in rural America. It is possible to make a loan to a credit worthy farmer borrower, knowing that the economy surrounding the borrower falls short of assuring him of the ability to repay that loan promptly when due. If the borrower defaults, the banker might accuse him of poor management, and the borrower might, in turn, accuse the banker of bad judgment for loaning him the money in the first place when we knew that farm prices are rotten, and farm net income is historically low. But, if neither of them know that farm prices are rotten, and farm net income is historically low, who is to blame? Most of the defaults on farm loans in the recent past, and today, are due to rotten farm prices. The blame lies with those who have the power to influence farm prices.

What began as a minor $6.4 billion correctable dislocation of national income in 1952, $4 billion of which was deducted from farm income, escalated into a single year loss of $148.7 billion realized net farm income by the end of 1982, and a very large $491.1 billion distortion in the national income share.

Both are now beyond any correction known to man. This changed not one iota with the adoption of "supply side" economics, and it won't change until parity is restored to an average of "100%" on all raw materials.

The cumulative effect of the dislocated income in all sectors of the economy has resulted in the creation of a total public and private debt of more than $12 trillion, for which there is no corresponding provision or ability to repay.

Is it any wonder why people who are put upon by such massive stupidity sometimes consider resorting to violence to make their point with society? Fortunately or unfortunately, there is very little comprehension of what we have done to American agriculture in the past forty years as a matter of deliberate government and public policy.

And there may be even less comprehension, if that is possible, of the irreparable damage that this misguided farm policy has inflicted on every other sector of the United States economy.

Name any single economic problem in the United States today, and it can be traced back to the Farm Act of 1953 and its predecessor, the Farm Act of 1948 when we "sovietized" the American agriculture economy in the name of economic freedom. What a tragic blunder this has turned out to be.

Earlier we stated that when the economy was in near balance, $1.00 of excessive debt would generate $1.00 of factory wages, and $5.00 of national income. However, that is no longer true. The severe distortion of national income share that has taken place since 1952, and the injection of $12 trillion of debt, has virtually eliminated any hope of restoring the economy again with more debt injection. Interest costs are higher than savings, therefore debt is off and running to infinity. The effort to stimulate the economy with new injections of debt may still be made, but it will result in rapidly rising interest rates and total monetary collapse in a short period of time.

Those who wait for that final inflationary spiral to get out of debt will more than likely miss the opportunity because it will happen so fast.

Debt is no longer even a temporary alternative to earned income and profits generated by an economy in historic equilibrium. It never was, of course, but it has taken forty years and one of the most appalling examples of fiscal and monetary mismanagement in human history to prove to a new generation of experts in the United States what the French nation learned so painfully exactly two hundred years ago.

What we must recognize is that economic equilibrium occurred in the United States from 1943 to 1952. It was brought about by a combination of circumstances which were carefully coordinated by government mandate, and WWII. The effect was (1) an average of "100%" of parity farm prices for the entire ten year period; (2) relative stability in prices and wages; (3) a steady flow of earned income which was adequate to purchase all of the goods and services available, for cash; (4) it provided sufficient savings with which to purchase enough government bonds to cover the cost of the war; (5) it paid for the costs of the post-war reconversion and balanced the federal budget in 1947, 1948, 1949 and 1951—a phenomenon which has occurred only one additional year since 1952, and then via statistical manipulation; (6) it maintained a sound dollar of consistent purchasing power throughout the period; (7) it sustained a solvent banking system and provided cheap interest rates; (8) it maintained approximate full employment with an average national income of only $213.8 billion per year. Why all of this remarkable economic performance? It happened because the economy of the United States was in equilibrium. It was in balance, and all sectors shared equitably at approximately the same share of national income percentagewise every year. The little imbalances that occurred from year to year were modest and self-correcting. It was a matter of distribution. Dollar totals mean nothing if there is maldistribution.

During the ten year period of 1943-1952, with an annual average national income of $213.9 billion, unemployment reached its highest level in 1949—a total of 3,367,000 workers. In 1983, with a national income of an estimated $2.5 trillion, nearly twelve times more money than during 1943-52 on an annual basis, unemployment was nearly three times greater. It isn't the number of dollars, but the precision of their distribution that is important. And the key to their distribution is dependent on economic equilibrium, and economic equilibrium is dependent (1) upon that amount of increase in raw material production needed annually to provide the needs and the wants of the modest increase in population; and vital to this factor is (2) the absolute necessity to price this raw material at a floor of 90% of parity in order to provide the cash to assure its consumption, without the injection of "excessive" debt.

We know that during 1943-52 the 90% of parity price floor resulted in an average return of "100%" for the ten year period. The reason for providing a 90% floor was to discipline the producer and to give the market place an important degree of control.

The experience in 1943-1952 was that prices seldom reached the support floor of 90%. Traders knew that prices could not go below 90% of parity, and generally came into the market at something near 90%. The producer, on the other hand, tended to curtail production of commodities selling at less than "100%" of parity, and switched to those commodities for which there was more demand which were also selling at "100%" of parity or above. Thus when products got into short supply and prices rose, farmers switched from lower priced products with less demand. This would cause the higher priced commodity to decline in price as production increased, and the cheaper commodities to rise in price as production increased, and the cheaper commodities to rise in price as production decreased. This is the classic example of "market discipline" as it is applied by the "law of supply and demand." However, it is controlled supply and demand that prevented wide swings of prices below 90% of parity, and could likewise prevent surges above 110% of parity. The value of this plan is that it limited disequilibrium to relatively modest changes in national income share, in an annual cycle, which is self-correcting from year to year and in the long run.

Congressman Harold Cooley from North Carolina, then the chairman of the House Committee on Agriculture in the United States Congress, made the following statement about the 1943-52 period of economic equilibrium in a speech on April 10, 1964.

"This old farm program worked when the great majority of farmers wanted it to work and were willing to cooperate.

"For eleven consecutive years prior to 1953, the average prices paid to farmers were at or above "100%" of parity with the rest of the economy. There was prosperity on the farms and along Main Street.

"Rural America—the countryside and Main Street—looked secure then and for all the years ahead.

"The government, with broad cooperation of farmers, supported the prices of the major storable crops for twenty years at an actual profit of $13 million. The government made this profit be selling commodities—wheat, corn, cotton, tobacco, rice and peanuts—taken over in price supporting operations.

"Those who deprecate the role of the farm program in this great era of farm prosperity emphasize that this period embraced war and postwar years, when the demands for the products of our farms were high, but they ignore the fact the markets—at home and abroad—for farm commodities have been greater the last ten years than during any other period of our history; and they forget that the farm economy collapsed after WWI, and this did not occur following WWII when the farm program was working.

"For eleven years—1942 through 1952—farmers had bargaining power in the marketplace. Supply and demand were in reasonable balance and farmer enjoyed price insurance through the farm program.

"But there was a turnabout—a flopover—on farm policy in Washington eleven years ago [1953]. The change came with the advent of the Eisenhower administration when Ezra Taft Benson became the Secretary of Agriculture. Production adjustment was deemphasized and price supports were lowered. This occurred at the very time broad strides in efficiency and ex-

TABLE 8
THE REAL STORY

Column 1. *That percentage share of National Income earned by agriculture annually as Realized Net Farm Income from 1929 to 1990*

Column 2. *The total loans of all commercial banks in the United States in ratio to total deposits from 1929 to 1990.*

	PERCENT OF NATIONAL INCOME	LOANS IN RATIO TO DEPOSITS
1929	7.2	68.9%
1933	6.3	47.2%
1939	6.2	28.3%
1940	5.5	27.2%
1941	6.2	30.4%
1942	7.4	21.6%
1943	7.1	18.1%
1944	6.6	16.9%
1945	6.9	17.4%
1946	8.3	22.4%
1947	7.8	26.4%
1948	8.0	29.7%
1949	5.9	29.6%
1950	5.8	33.7%
1951	5.9	35.0%
1952	5.1	36.6%
1953	4.2	38.3%
1954	4.4	38.2%
1955	3.4	43.0%
1956	2.1	45.7%
1957	3.0	46.6%
1958	3.5	45.5%
1959	2.6	50.4%
1960	2.7	51.2%
1961	2.7	50.2%
1962	2.6	53.5%
1963	2.4	56.7%
1964	2.0	57.2%
1965	2.2	60.7%
1966	2.2	61.8%
1967	1.9	61.1%
1969	1.7	67.9%
1970	1.8	65.2%
1971	1.7	64.5%

TABLE 8
(continued)

	PERCENT OF NATIONAL INCOME	LOANS IN RATIO TO DEPOSITS
1972	2.0	67.3%
1973	3.0	72.6%
1974	2.3	
1975	2.0	73.4%
1976	1.4	69.5%
1977	1.3	71.0%
1978	1.5	72.4%
1979	1.5	77.6%
1980	.9	83.5%
1981	1.3	81.2%
1982	1.0	86.5%
1983	.5	85.8%
1984	1.0	89.7
1985	.9	91.2
1986	1.0	94.5
1987	1.1	94.5
1988	1.0	95.5
1989	1.2	98.2*
1990	under revision	

Commercial Bank Data, Federal Reserve Bank of St. Louis

*This represents an all-time high.

All data presented by the Federal Reserve are revised constantly, therefore Rossiter's earlier "The Real Story" tables will vary slightly from this updated and most recent version.

plosions of production and lowering price supports prevailed to a marked degree.

"Prices came down. Great surpluses accumulated. Makeshift programs for the individual crops, for the most part, now hold farm income up to some degree, but at great costs.

"In the ten years—1953 through 1962 inclusive—while all other segments of the economy have been booming, the net income of agriculture has been $25 billion less than in the previous ten years—1943-1952. Meanwhile, the Department of Agriculture spent, for all purposes in those ten years, $35 billion more than in the previous ten years. The costs from 1953 through 1962 were almost $20 billion more than all expenditures of the Department in the previous ninety years of its history.

"Many farmers have turned against their own program—the program that prevailed during the years of our greatest era of farm prosperity. Why, and for what reason, I shall never understand.

"Farmers have lost bargaining power in the marketplaces, and "100%" of parity for agriculture—generally approved and accepted by the public a decade ago—is hardly more than a dream today."

The last half century is touted to be the longest period of uninterrupted prosperity in the history of the United States. It is seen as a period in which the national income of the United States has increased from *$42.8* billion in 1932, to an estimated *$2,544.4* billion in 1982—nearly *fifty-eight* times more than in 1932. Realized net farm income, in the same period, increased from $2.1 billion in 1932, to an estimated $19.2 billion in 1982—not quite ten times more than 1932.

This tells you that agriculture is still doing the same job that it did fifty years ago—providing the raw materials for feeding, clothing, and housing society—but it is receiving only one-sixth the increase in national income share that the other sectors in the economy have received. Agriculture's loss of national income share in the last forty years of unprecedented national prosperity is nearly three times greater than it was from 1929 to 1932—the worst national depression in history.

Such incredible statistics should inspire the sympathy, if not the understanding, of the meanest bankers in the country.

I'm inclined to believe that the figures in Table 8 on pages 130-131, are the most significant of any in this series of exhibits because it consolidates the cumulative totals so concisely.

For example, the cumulative total of all the annual losses of cash that agriculture failed to earn as realized net income on production that was sold during the years—had the industry's share of national income remained consistently at 6.63% per annum—would not only have paid off every cent of the debt owed by every farmer in the United States, but there would have been enough left over to buy and pay for every one of the dollars invested in the agricultural economy, lock, stock and barrel. And there would still have been enough money left for celebration of the farmer's good fortune.

However, had agriculture earned 6.63% of national income consistently after 1952, it wouldn't have owed billions in 1990, nor would the farmer need to have purchased all of the assets of the agriculture economy because he already owned that, and it would have been virtually debt free.

The fantasy that the nation is suffering from is the idea that it is possible to avoid profits and earned income, and inject trillions of debt into the economy without doing irreparable and permanent damage to the economy, which will cause problems for the next hundred years or more.

The fantasy is, that we in this generation feel that we have a right to obligate trillions of the income to be earned by future generations, which will have to be repaid by not just our children, but also our children's children, and very likely their children's children. Debts must and will be repaid one way or another. One of the lessons we bankers learned in the 1930s is that "if the debtor doesn't pay, the creditor will."

The fantasy is that we in this generation have been perfectly willing to obligate the income of future generations—who have no say in the matter—and probably expect them to remember us with love and affection when we are gone.

The fantasy is that we have a right to use precious resources —which we could not afford without an injection of excessive debt—which we should be holding in reserve for future generations as their birthright.

Christian Science Monitor carried a message damning in its finality: "As Wharton economist Jan Vanous recently noted, the morass of Soviet agriculture has created a situation in which it is far more profitable for the Soviets to buy grain rather than produce it themselves. Grain imports . . . are calculated to have saved the Soviet economy . . . billion[s] over the cost of domestic production."

In the meantime farm leaders from the United States, and representatives of the USDA are concentrating on persuading Common Market countries to reduce their support prices to equate the reduced market price in the United States.

If this primer means anything, it is that there is a legitimate way to turn the economy around. If we refuse we will become the victims, as well as the agents, of our own destruction.

AFTERWORD

Long reliant upon raw-material prices that are too low to support a true prosperity, the American economy cannot restore itself through free market forces alone. This is because the free market system today does not function due to the concentrated power of raw material marketing conglomerates. In order to put the economy back on the right track we must enact a new law.

The proposed National Economic Stability Act (NESA) is such a law. Simply stated, this proposed legislation would establish a price structure for basic storable commodities harvested off 75 to 80% of America's tillable acres. It would also regulate the parity values at which semi-storable commodities—potatoes, meat and dairy animal products—could enter trade channels. Once this production is priced correctly, values of perishable produce and other raw materials would fall in line, laying the foundation for stability and prosperity unlimited.

1. NESA eliminates current federal crop subsidy, financing, and commodity buying programs. Since NESA will establish a new and truly fair market, continued government intervention will be unnecessary. Government may purchase food for such programs as foreign aid and school lunches, but it must buy all

such commodities on the open market and pay the same price as other buyers.

2. Prices for storable foodstuffs and fiber (corn, wheat, barley, oats, soybeans, rice, cotton, potatoes, milo) and for livestock, dairy animals and poultry will be set at a minimum of 90% of parity and a maximum of 115% of parity. These prices will be based on USDA figures drawn from the 1910-1914 base period, and they may fluctuate within the limits as the market demands. To ensure uniformity, the Consumer Price Index and the Wholesale Price Index will also be reset to fit the parity price structure.

3. The minimum wage equals the parity price of a bushel of corn. This feature ensures the adjustment of the whole economy to the new price structure for primary supply products, and it guarantees mass purchasing power sufficient to absorb the supply of those products at the new market prices.

4. Agricultural producers may plant the crops they wish in the amount they wish, but they can sell only what the domestic and export markets will absorb. Production-for-sale quotas for each producer will be authorized by marketing certificates issued annually.

5. Excess product (i.e., over the amount allowed by the marketing certificate) must be stored at the producer's expense. The stored product will provide an important hedge against bad weather and crop failure. Also, it can provide the basis for establishing a national food and fiber reserve, which would greatly strengthen the nation's defense. This provision distinguishes NESA from the current parity pricing law for milk, which encourages overproduction by obligating the government to buy excess production no matter how much it exceeds actual demand.

6. Imports of the products governed by NESA must be priced domestically at 110% parity. This provision is needed to prevent the domestic market from being flooded by cheap commodities from abroad. Ultimately, too, NESA will work to the benefit of raw materials producers overseas, by guaranteeing them a fair and adequate price for their products and thereby

establishing their economies on the firm foundation needed for sound economic development.

Implementing NESA requires only a simple organizational structure, one that gives agricultural producers a major voice and that trims governmental involvement to the minimum necessary to ensure efficient and impartial administration.

Prices for regulated commodities will be set monthly for each Farm Credit District by the USDA under the supervision of the National Board of Producers. The Board, chaired by the Secretary of Agriculture, will be elected by producers of the covered commodities. Each Farm Credit District will elect one member. Every county and state agriculture department will be advised by similar producer boards comprising three regular delegates and two alternates, all of whom must be producers of one or more basic commodities.

It will be the responsibility of the National Board to establish domestic and foreign demand before the production season begins and to update these figures every quarter. The Board will also issue marketing certificates and update them quarterly. Certificates will be based on the production history of each grower, with allowances made for new and retiring producers as needed. If the National Board finds that demand falls below select year production levels, the shares of larger than average producers will be fractioned downward.

To ensure compliance with NESA's regulations, sellers and buyers must report first-entry-to-market transactions to the county ASCS office on a monthly basis. This requirement also covers co-ops, which are allowed to store, process, and market all members' produce covered by marketing certificates. Covered production may also be used as collateral for loans.

Contractors who buy or sell at prices above or below the set levels will be fined three times the dollar amount of the transaction. Fraud, including selling without a marketing certificate, will be subject to other legal penalties.

A commodity tax will be levied on each covered commodity upon first entry to the market. The amount of the tax will be computed to cover the cost of administering the program plus commodity marketing promotion and production research. The tax is to be included as an expense in the parity formula, so that domestic and foreign consumers and buyers will be contributing to the programs.

In keeping with NESA's self-sufficiency, all CCC loans will be transferred to private banks and financial institutions, and all federal programs governing the commodities regulated by NESA will be suspended. Likewise, special concessions granted to corporations engaged in productive agriculture will be suspended. In short, agricultural subsidies for the affected commodities will end.

If this economic policy is to succeed, it must aim at the root cause of the current malaise: primary supply price depression. A new policy focus is needed.

The health of the primary supply sector determines the health of the whole economy. Revitalizing the primary supply means increasing net income. And that means equitable raw materials prices.

GLOSSARY

GLOSSARY

CANTILLON, RICHARD. He wrote *Essai sur la Nature du Commerce in General* in 1755, which included that famous passage, "The land is the source or matter from whence all wealth is produced. The labour of man is the form which produces it; and wealth in itself is nothing but the maintenance, conveniencies, and superfluities of life." This single thought made him the forerunner of the study of economics, and helped shape the thinking of the Physiocrats. Cantillon would have chortled at the idea that wealth could be conjured up out of thin air.

CAREY, HENRY C. An economist who saw individuals as molecules in society; indeed, he lifted his level of abstraction high enough to discern that agriculture needed the counterbalance of healthy manufactures in order to have convenient markets. He warned against America becoming a great farm for England. He saw British policy as "selfish and repulsive, its essential object being the separation of the consumers and producers of the world. In that direction lie poverty and slavery." He hated cheapness because a few become wealthy, but the people generally remain poor and wretched. He ruled out free international trade as a viable course for a nation interested in its people first.

CLASSICAL SCHOOL. Generally, a school of economic thought spawned by Adam Smith via his *Theory of Moral Sentiments* and *An Inquiry into the Nature and Causes of the Wealth of Nations*. Although Adam Smith stated many valid ideas—including the concept that agriculture came first—others, namely David Ricardo, Thomas R. Malthus, John Stuart Mill, Jean Baptiste Say—became enchanted with "laws" that ignored institutional arrangements and substituted business principles for economic principles. This meant economic affairs settled themselves best for everyone when man made no attempt to regulate. Terms like orthodox school, Manchester school, individualist school, etc., are sometimes used. None comprehend the role of raw materials entirely. Say's Law of Markets came close, but never discerned the parity requirement.

COMMITTEE FOR ECONOMIC DEVELOPMENT. Founded in 1942, it is a spinoff from Council on Foreign Relations. It crafted the blueprint for debilitating agriculture and annihilating the parity concept. On July 15, 1962, a day that will live in infamy, CED issued its *An Adaptive Program for Agriculture*.

COMMONS, JOHN R. A Wisconsin-based institutional economist who joined Walton Hamilton in the view that "economic theory must be based upon an acceptable theory of human behavior." He did not have an explanation for how purchasing power was generated in the first place. Some of his students figured it out. We call it raw materials economics.

COMPARATIVE ADVANTAGE. The ability of one nation to produce cheaper than the next. This advantage is usually based on wage differentials, sometimes on natural resources. A dime an hour wage country has an advantage over a $10 an hour wage country, capitalization in either case being in the hands of internationalists.

CONJECTURAL ECONOMICS. A term used by the author of this primer. It describes most economics because almost all schools are based on suppositions which do not exist as long as there are wars and differences of nationality, and institutional arrangements take advantage of each other.

COULTER, JOHN LEE. Chief economist for the U.S. Tariff Commission at the time of the McNary-Haugen bill fights. He was a schoolman, Dean of West Virginia College of Agriculture, President of North Dakota A & M. He served on the editorial staff of the *Quarterly Journal of the American Statistics Association* and the *American Economic Review*. He taught Carl Wilken how the laws of physics governed raw materials economics.

COUNCIL ON FOREIGN RELATIONS. Founded in 1921, this sub-rosa government of the United States did not attain status and power until the Rockefeller, Ford and Carnegie Foundations poured in money. CFR offered the first of its policy papers in 1939. During the Truman administration, CFR members literally seized the State Department. In 1945 no less than 40 CFR members were on hand to help form the United Nations Charter. Among them were Alger Hiss, Edward R. Stettinius, Leo Pasvolsky, John Foster Dulles, John J. McCloy, Nelson A. Rockefeller, Adali Stevenson, Ralph J. Bunche and Tom Finletter. These names are recited here to illustrate to what extent the American State Department had become the fief of a policy writing group named Council on Foreign Relations. By the time John F. Kennedy was staffing his administration, according to political historian Theodore H. White, writing in *The Making of a President 1964*, sixty-three of the first eighty-two names submitted for top government posts were CFR members.

DIVISION OF LABOR. Generally, the assignment of duties on an assembly line so that the joint effort becomes more productive than the individual effort might be. The term can also be used to define changes in economic function, such as when farming chores (the slaughter of animals, for instance) are divided from the farm and made a separate enterprise.

ECONOMIC INDICATORS. A statistical publication prepared for the Joint Economic Committee by the Council of Economic Advisers. It provides monthly figures on national income, compensation of employees, farm income, nonfarm proprietors' income, rental income, corporate profits and capital costs (interest), among other data.

ECONOMIC REPORT OF THE PRESIDENT. An economic report mandated by the Employment Act of 1946. As supervised by the Council of Economic Advisers, it details public policy and provides figures that codify data in the *Economic Indicators.* Lately, the *Economic Report of the President* has been used to hard-sell GATT, free international trade, a debilitating farm policy, and theory period conjectural economics now common fare in most universities aligned with the military-industrial complex.

ECONOMY. A term used to describe the several actions and processes having to do with the creation of goods and services to answer human wants.

FACTORS OF PRODUCTION. A term generally used to define the inputs used to produce. Three common heads used are land, labor and capital. A fourth factor of production often cited is entrepreneurship.

FORDNEY-McCUMBER TARIFF. Passed in 1922, it answered some of the problems set up by discovery of the parity formula a few years earlier. Fordney-McCumber raised tariffs and established the concept of flexible tariffs, chiefly for industrial goods.

FORTUNE 500. A list of 500 firms designated by *Fortune* magazine as the leaders in industrial America.

FREE TRADE. The absence of export and import duties in international trade.

FUTURE ADVENTURES OF ROBINSON CRUSOE, THE. A comic book issued in England about a quarter century ago for the purpose of explaining the merits of a gold standard. Some of the panels have been used in this primer because the story has often been used to illustrate the first stage of economic growth.

GOUGE, WILLIAM M. The author of *A Short History of Paper Money and Banking in the United States to Which is Prefixed an Inquiry into the Principles of the System.* Writing during the Jackson era, Gouge took the flim-flam out of banking. Everything he wrote about was eclipsed by the arrival of the Federal Reserve System. The Fed was to keep a steady flow of sound credit going out to the nation; help economic growth; provide for a stable dollar; and keep a positive balance in the system of overseas payments. It has accomplished none of the above. Parity prices can and have done everything the Fed has failed to do. In fact, par values for raw materials is the only antidote for wild credit. When Great Plains farmers last enjoyed full parity, many a local liquor store made more money than the local bank. And the Fed found its powers for mischief emasculated. Gouge would have predicted as much.

GROSS NATIONAL PRODUCT. The computed value of goods and services before deduction of depreciation charges. It includes personal consumption expenditures, meaning goods, services and income in kind. GNP also includes new homes, domestic investments and inventory changes, net foreign investments and government purchases of goods and services, including killing machines. Generally excluded are subsidy payments. A better term would be gross national expenditure. Raw materials economics prefers to use a more exact concept—gross national income.

HOLLIS, CHRISTOPHER. A British economist, he authored *The Breakdown of Money*, published in 1937. His seminal work greatly influenced Carl Wilken, John Lee Coulter and Charles Ray. The principles enumerated in his book helped provide a foundation for the stability equation invoked to fight WWII. Its last chapters pointed to the fallacy of international loans.

INSTITUTIONAL ECONOMISTS. This school of economics places great emphasis on the social environment. Thorstein Veblen (1857-1929), for instance, liked to beat up on the idea that an economy could override the warren of foxes that managed business. Their greed, prestige and power were bound to overshadow the notions of supply and demand and even good government. The business of monopoly, the transfer of wealth from one group to a more powerful group, all conspired to undo the norms society sought. The problem with raw materials economics is that there are not enough men of good will to invoke its "reason." Much of the reasoning behind Carl Wilken's raw materials equation was harvested from the institutional economics of Wesley Clair Mitchell and John R. Commons via the scholarship of John Lee Coulter.

INTEREST. There are many theories of interest. Indeed, whole books weigh down library shelves dealing with the subject. The old classical school considered interest much like rent on money, governed by supply and demand. Actually interest is usually governed by monetary authorities and institutional arrangements. In our time, faltering parity has been used to short-circuit returns and open up a market for lenders. The rate of interest by trapped borrowers is merely a lug, and not based on the esoteric calculus proposed by E. von Böhm-Bawerk, John Maynard Keynes, or Irving Fisher. K. Wicksell may have had a point with his natural rate of interest. Generally, the natural rate of interest is approximately 3% plus the rate of inflation.

IRON LAW OF WAGES. A theory formulated by Ferdinand Lassalle to the effect that wages tend to equal what a worker needs to maintain subsistence. Lassalle held that when wages increased beyond a bare minimum, more workers would bid to drive it down. Lassalle's iron law came out of the German socialist movement during the last part of the nineteenth century. It is also called the *brazen law of wages* and the *subsistence laws of wages*.

JEVONS, W. STANLEY. He authored a marginal utility theory and was a founder of econometrics. He conceived the idea of moving averages and detailed a trade cycle theory based on sunspots. His contribution is recalled with each double sunspot maxima, such as the one in October 1929, and the one projected for October 1991.

KEYNESIAN ECONOMICS. A body of economic thought named after the British economist, John Maynard Keynes (1883-1946). According to Keynes, savings must be offset by investment or unemployment will result. Keynesian thought had it that people save according to their propensity to consume and tend to hold onto liquidity, for which reason savings may not be mobilized. In this event, it is up to the government to spend and regulate the rate of investment in order to assure full employment. A major work by Keynes is *The General Theory of Employment, Interest and Money*. Keynes was not oblivious to the generating power of agriculture. He noted that on food produced by the farmer, the price paid for it became the income of those in the business, and thus increased their expenditures. A higher living cost created pressure for higher wage costs. Thus prices rose all around, and an inflationary spiral set in. Thus, he held, the price paid out at the farm raw materials level increased effective demand and created a trend to inflation. Keynes thus arrived at and ignored the solution embodied in raw materials economics. Non-inflationary development could be achieved only when quick-yielding investment was made in the consumer goods sector, principally agriculture, because agriculture generated the surplus melded to support capital developments on an earned basis. Keynes opted to create credit and to subject the economy to all the evils that credit and inflation would visit upon it. Raw materials economics argues that progress can be more dramatic that way. Gross National Product can dance a dizzy tune on its way up—but in the end the laws of physics have to be obeyed.

KUZNETS, SIMON, 1901-1990. A pioneer in the study of econometrics. He developed the *Economic Indicators* for President Roosevelt in the 1930s. In later years he was chagrined to learn that economists were using his indicator figures for the wrong purpose, and not for long range trend predictions. He was a pioneer in making major studies of income distribution. Carl Wilken and his associates relied on the works of Kuznets for primary data with which to make the connection between raw materials income and national income.

LABOR. One of the classic factors of production. Economic texts deal with labor under many heads—labor-management (Taft-Hartley Act), labor piracy, labor theory of value, etc.

LAISSEZ-FAIRE (LAISSER FAIRE, LAISSEZ PASSER). Literally, let things proceed without interference by government, but cartels, monopolies, institutions with clout, should all remain free to interfere. Adam Smith applied *laissez faire* to foreign trade. He advocated the withdrawal of restrictions imposed by mercantilist interests, meaning the crown's trader-government complex.

LAUDERDALE, LORD, 1759-1830. A Scotchman who opposed Adam Smith because he was confused about the difference between individual and national wealth. His dictum: individuals seem generally to grow rich by grasping a portion of existing wealth; nations, by the production of new wealth. Unfortunately, governments too often accommodate the wishes of wealthy individuals and corporations are hard on the hunt for more wealth.

LIST, FREDERICK. An economist who loomed large on the American scene as a foe of import invasion. He was a German who published *National Zeitung* at Harrisburg, Pennsylvania. His chief aim in life was the overthrow of the "School," as he called Adam Smith and his followers.

MACRO-ECONOMICS. An overview of the economic scene as opposed to examination of details that may not serve up comprehension of the whole. For instance, a study of increments of bid increases at a bull auction will do practically nothing to uncover an understanding of why there has to be par exchange between the several sectors of the economy. The statistics in the *Economic Indicators*, designed by Simon Kuznets, furnish data needed to understand the macro-scene.

MALTHUS, THOMAS ROBERT 1766-1834. A minister by trade, Malthus is best remembered for his essays on population. He held that the laws of nature permitted only an arithmetic increase in the food supply, but man's propensities brought on a geometric increase in the human population. The result, he saw, would be gloomy. Malthus nevertheless saw that a failure to distribute income would result in underconsumption.

MARGINAL EFFICIENCY OF CAPITAL. This is a statistical term used to denote the difference in earnings by an asset and the cost of reproducing that asset. Unfortunately this type of dead reckoning often fails the test of reality in the real world.

McNARY-HAUGEN BILLS. Passed twice by Congress, McNary-Haugen was vetoed twice by Calvin Coolidge. The McNary-Haugen measures sought to preserve parity for agriculture by preventing import invasion.

MILL, JOHN STUART. Probably the king of classical economists. Mill illustrated how *price* was determined by the equality of *demand* and *supply*. He dealt with elasticity of demand, even though Alfred Marshall later received credit. The gist of his thinking was that distribution was properly an institutional consideration.

NATIONAL INCOME. A measure of the total factor cost of all final goods and services produced in the economy during a specific time frame, except those emanating from illegal activities and the work of housewives.

NORM (NATIONAL ORGANIZATION FOR RAW MATERIALS). A small think-tank composed of members who hold that monetization of raw materials on par with wages and capital costs holds the key to full employment, stability and structural balance for the economy.

NET FARM INCOME. This is computed by deducting total production expenses from the sum of cash receipts generated by the sale of farm products, receipts from government sources, value of farm products consumed on the farm, rental values of farm homes, and the value of any change in the physical of the physical volume of farm inventories of livestock and crops.

PARITY. The general concept of making a comparison between prices farmers receive and those they pay. It was developed by Professor George M. Warren, and popularized by George Peek of the Moline Plow Company. The general idea was published as a paper by Warren in 1921 and reprinted as USDA Bulletin 999, *Prices of Farm Products in the United States.* Warren tested his concepts in open forums. He presented varicolored chart showing the separate price movements of twenty farm commodities, the weighted average price for thirty-one farm products, and the movement of the all-commodities index of the Bureau of Labor Statistics. Warren's figures revealed the disparity between those all-commodities figures and what the farmer was getting. Parity was written into law in the 1930s, 1910-1914 being base period "100." Parity revealed that unless farm prices reflected averages of all-commodities for the crop year to follow, buying power would be short-circuited. See Lesson 8.

PHYSIOCRATS. A school of economics characterized by specific works and ideas, such as the *Tableau Economique*, by Quesnáy and *Reflexions* by Turgot, etc. It was Turgot who held that agriculture was the chief source of wealth and that manufactures and trade depend on it. *Reflexions* is still read as a milestone in economic thought. Unfortunately the creation of wealth from pure smoke or thin air holds economic thinking in thrall these days.

PROFIT. Generally, the return on investment. The term is usually associated with a business equation, and the business operator often sees profits as predatory, or achieved at the expense of another. Profit for the economy can come only from the raw materials of the planet, because the equation is man debited, nature credited. The value man chooses to put on raw materials determines the profit all others in the economy must share. The raw materials exchange equation is based on this premise. All other transactions in an economy are man debited, man credited—a wash.

PROPENSITY TO CONSUME. This is a Keynesian term—P = C/Y. P = propensity to consume, C = consumer expenditure, Y = income. The result is a percentage, and the propensity to consume is a percentage figure.

RAW MATERIALS ECONOMICS. That body of economic thought that identifies the raw materials of the earth as the foundation materials for economic growth. Raw materials economics (RME) holds that monetization of raw materials (production times price) determines the profits and savings at the end of the cycle. In this system, parity—defined as an index of prices down track after raw materials harvest—must start the next cycle. RME holds that Say's Law of Markets and supply and demand cannot serve the greater good because institutional arrangements intervene to serve there own end, and therefore government must mandate raw materials parity if periodic inflation, deflation and convulsion are to be avoided, and full employment served.

RAY, CHARLES B. An engineer who worked with General Wood at the Panama Canal, later with the Raw Materials National Council. His forte was raw materials economics. He was an associate of Carl H. Wilken.

RENT. The income to a person who holds legal title to a durable good such as an acre of land, a piece of real estate or a piece of equipment.

REVERSE MULTIPLIER. The raw materials economics principle that a shortfall in income due to underpricing of raw materials has a reverse multiplier effect that denies generation of national earned income according to the ratio dictated by the state of the arts.

RICARDO, DAVID 1772-1823. A stockbroker who made a name for himself in the great economic debates of the late eighteenth and early nineteenth century. His treatises on rent and land have been swept aside today because agriculture is no longer seen as a sector of consequence. His name is often associated, erroneously, with the iron law of wages. Ricardo's thinking suffered became of a paucity of good statistics.

SAVINGS EQUALS INVESTMENT. This worried the early economists tremendously, since too much saving short-circuited markets exactly the way faltering parity does the same thing. Ever since the Wilson administration, the institutional arrangements for finance have learned how to mobilize everything excepted hoarded funds. Moreover, investments are now made on the basis of debt.

SAY'S LAW OF MARKETS. Say's law said that the total supply of economic goods necessarily equaled the total demand. The idea here is that goods really exchange for goods, money being no more than a medium of exchange. Therefore all goods represented both a supply and a demand. In a macro-sense, Jean Baptiste Say (1767-1832) was somewhat on track, but he missed the principle of faltering parity short-circuiting buying power, for which reason supply and demand cannot be the same. Buying power cannot be manufactured by a disparity between the several sectors of the economy.

SCHOLASTICISM. Originally this term meant schoolmen of the 13th century, especially those who framed their logic with esoteric arguments. As used today it means reliance on nitpicking arguments that would offend uncommon good sense if presented in terms most people could understand.

SISMONDI, JEAN CHARLES 1773-1842. A Swiss historian, he argued against *laissez faire* and for state intervention in economic affairs. He understood that consumers without money could not buy, and thus headed schemes to pension workers, provide sick pay, etc. He joined Malthus in attacking Ricardo for remaining blind to the fact of underconsumption.

SLIDING PARITY. As written into law in 1948, sliding parity meant the annihilation of the parity concept. First, the 1948 farm act set up base periods "100" as moveable feasts, so to speak. Each decade "100" was moved ahead regardless of conditions. A 60 to 90% formula was established. This meant that farm prices would be maintained at between 60 and 90% of the norm sought by George Warren. See Lesson 8.

SMITH, ADAM 1723-1790. Basically, Adam Smith rejected the view of the physiocrats that agriculture and extractive industries were prime movers. He hung his hat on maxims—that a hidden hand guided mankind to greater welfare, that free competition was the one essential ingredient each economy required. He was more a sociologist than an economist. He bowed to labor, but blessed the idea of market supply and demand. His basic failure was his inability to take the overview needed to understand the total mechanism. As a consequence, Adam Smith lost himself in micro-equations and then attempted to extrapolate his finding to being the norm on the macro-scene. See CLASSICAL SCHOOL.

STAGES OF ECONOMIC GROWTH. A well advertised theory by Walt Rostow to the effect that economic societies have five stages: as a traditional society, slow and plodding; as a launching pad society, ready to take off; as one in a take off mode, with resistance to progress overcome; as a drive to mature society; and finally an era of high mass production and sustain ability. Rostow's book *The Stages of Economic Growth* is mentioned here because it is typical of fond-hopes economics. Its general thesis was lifted from Friedrich List, a German economist who taught the Founding Fathers that government ought to aid in the transitions from one stage to the next. Rostow nails down none of the realities covered by raw materials economics, and ignores the points List made part in his original presentation.

STATE OF THE ARTS. This refers to the level of technology achieved in a society. It also implies that technology cannot be withheld from one sector of an economy and stay on to benefit another. If new technology arrives—much like an added bucket of milk in a tanker—it blends quickly. Thus there is little reason to hold that technology for agriculture has outrun technology in industry, and therefore should reflect lower prices for commodities.

SUPPLY AND DEMAND. The business concept that price varies directly, but not necessarily proportionally, with demand, and inversely—also not necessarily proportionately—with supply. In terms of macro-economics there really is no such thing as supply and demand.

TRADE TURN. Essentially a multiplier. As used by Carl H. Wilken, it meant the ratio at which raw materials income "turned" through the trades to become national income. Trade turn, computed off either farm income or the total of raw materials income, is governed by the state of the arts.

UNFORGIVEN. A book by Charles Walters Jr. that covers the life and thoughts of Carl H. Wilken. It is also a biography of the raw materials idea.

WILKEN, CARL H. The last survivor of the triumverate that made up the backbone of the Raw Materials National Council. The others were Charles B. Ray and John Lee Coulter. Wilken's story—and abstracts of his work—fill the pages of *Unforgiven*.

INDEX

INDEX

Glossary terms (page 139-156) are presented in small capital letters in this index and when first used in the text of this book.